The Electronics Industry Research Series

- The Taiwan Electronics Industry
- The Singapore and Malaysia Electronics Industries
- The Korean Electronics Industry

THE Singapore AND Malaysia ELECTRONICS INDUSTRIES

Donald Beane
Anand Shukla
Michael Pecht

CRC Press
Boca Raton   New York

Acquiring Editor: *Norm Stanton*
Senior Project Editor: *Susan Fox*
Cover Design: *Denise Craig*
Prepress: *Gary Bennett*
Marketing Manager: *Susie Carlisle*
Direct Marketing Manager: *Becky McEldowney*

Library of Congress Cataloging-in-Publication Data

Catalog record is available from the Library of Congress.

 This book contains information obtained from authentic and highly regarded sources. Reprinted material is quoted with permission, and sources are indicated. A wide variety of references are listed. Reasonable efforts have been made to publish reliable data and information, but the author and the publisher cannot assume responsibility for the validity of all materials or for the consequences of their use.

 Neither this book nor any part may be reproduced or transmitted in any form or by any means, electronic or mechanical, including photocopying, microfilming, and recording, or by any information storage or retrieval system, without prior permission in writing from the publisher.

 CRC Press LLC's consent does not extend to copying for general distribution, for promotion, for creating new works, or for resale. Specific permission must be obtained in writing from CRC Press LLC for such copying.

 Direct all inquiries to CRC Press LLC, 2000 Corporate Blvd., N.W., Boca Raton, Florida 33431.

© 1997 by CRC Press LLC

No claim to original U.S. Government works
International Standard Book Number 0-8493-3171-4
Printed in the United States of America 1 2 3 4 5 6 7 8 9 0
Printed on acid-free paper

CONTENTS

Chapter 1 Introduction 1
    1.1. History 1
    1.2. Geography 2
    1.3. Demographics 3
    1.4. Economy 4
    1.5. Education 5
    1.6. Culture 5
    1.7. Summary 6

Chapter 2 The Electronic Industry's Role in Singapore's Economy 9
    2.1 Background 9
    2.2 Status of the Industry 12
    2.3 Economic Indicators 13
    2.4 Manufacturing in Singapore's Economy 13
    2.5 Electronics in Singapore's Economy 16
    2.6 Role of Trade 29

Chapter 3 The Electronic Industry's Role in Malaysia's Economy 33
    3.1 Background 33
    3.2 Status of the Industry 35
    3.3 Economic Indicators 38
    3.4 Electronics in Malaysia's Economy 39
    3.5 Role of Trade 40

Chapter 4 Electronics Production in Singapore 45
    4.1 Semiconductors and Packaging 46
    4.2 Products and Services 49
    4.3 Case Studies 59

Chapter 5 Electronics Production in Malaysia 63
    5.1 Semiconductors and Packaging 63
    5.2 Products and Services 64
    5.3 Case Studies 67

Chapter 6 Government and University Support of Advanced Technology Development 71
    6.1 Singapore 72
    6.2 Malaysia 77
    6.3 Summary 83

Chapter 7 Conclusions 85
    7.1 Singapore 86
    7.2 Malaysia 93
    7.3 Summary 99

References 101

Index 103

PREFACE

Since Singapore was separated from the Malaysian Confederation in 1965 and Malaysia underwent political unrest in the late 1960s, both countries have made remarkable progress towards modernization. The new focus has been on developing an industrial, export-oriented economic base, a major portion of which is focused on the electronics industry. Today, Singapore and Malaysia own a significant share of the world's electronics market, conduct state-of-the-art research and development projects, establish foreign ventures, and support universities and research institutes devoted to producing the countries' future engineers, scientists, and technical leaders.

This book documents the technologies, manufacturing procedures, capabilities, and infrastructure that have enabled these countries to be so successful. This knowledge, coupled with knowledge of the planned future directions of Singapore and Malaysia's electronics industries, is vital for determining an appropriate competitive response for the U.S. An understanding of the present and future of Singapore and Malaysia's electronics industries is needed in order to determine in which market sectors to compete and in which areas subcontracting, outsourcing, partnership agreements, and foreign direct investment would be beneficial.

The electronics industries of both Singapore and Malaysia are described in this one book rather than in separate books because of the strong relationship between the two countries. Both countries are active members of the Association of Southeast Asian Nations (ASEAN). Water and other natural resources are carefully traded, as are human resources. For example, in 1995, Singapore employed 450,000 Malaysians. The two countries have also undertaken joint development projects that have affected their industrial and electronics development. Because of Singapore's acute labor shortage, Malaysia's educational and developmental needs, and the desire of Singapore to create a geographically-based development zone, these economic interactions are likely to accelerate.

This book consists of seven chapters. The first chapter provides a background to the social-cultural-political evolution and development that has led each country to its current status. Chapters two and three individually examine the impact of each country's electronics industry upon their economies. Chapters four and five individually examine each country's electronics industry. Chapter six provides an analysis of the relationships between each country's government, university, and industrial sectors. The final chapter gives a summation of our findings.

Research for this book was supported in part through a MANTECH program grant No. 60NANB500060 administered out of NIST, in part through the CALCE EPRC of University of Maryland, and in part through the NSF WTEC program managed out of Loyola College in Maryland.

Many people have supplied us with technical information and other resource materials, and we are extremely grateful to them. These people include the WTEC team of experts including Michael Pecht of University of Maryland, C.W. Wong of Texas Instruments, Joseph Ranieri of Universal Instruments, Amelendu Sanyal of Digital Equipment Corporation, Bill Tucker of IBM, and Sam Wennberg of Delco Electronics and Dr. Swan-Jin Beh of Singapore's Economic Development Board (Washington D.C. office).

ACRONYMS

AFM	Atomic force microscopy
ASEAN	Association of South East Asian Nations
ASIC	Application specific integrated circuit
AV	Audio-visual
BGA	Ball grid array
CAAG	Compound average annual growth
CAD	Computer-aided design
CAM	Computer-aided manufacturing
CD	Compact disk
CD-ROM	Compact disk read-only memory
CICFAR	Center for Integrated Circuit Failure Analytic and Reliability
CIM	Computer-integrated manufacturing
CL	Cathodoluminescence
CMOS	Complementary metal oxide semiconductor
COB	Chip-on-board
COE	Center for Optoelectronics
CPT	Color picture tube
CRT	Cathode ray tube
CWC	Center for Wireless Communications [Singapore]
DIP	Dual in-line package
DRAM	Dynamic random access memory
EBIC	Electron beam induced current
EDB	Economic Development Board [Singapore]
EDP	Electronic data processing
EPU	Economic production units
FDI	Foreign direct investment
FTZ	Free Trade Zone
GATT	General Agreement on Tariffs and Trades
GDP	Gross Domestic Product
GNP	Gross National Product
GSP	General System of Preferences
GVA	Gross value added
HP	Hewlett-Packard
IC	Integrated circuits
ICT	Integrated circuit test
IME	Institute of Microelectronics [Singapore]
IMF	International Monetary Fund
IMP	Industrial Master Plan [Malaysia]
IMS	International Microelectronics and Systems

IPO	International procurement office
ISO	International Standards Organization
ITAF	Industrial Technical Assistance Fund
JICA	Japan International Cooperation Agency
IR	Infrared
IT	Information technology
KHTP	Kulim Hi-Tech Industrial Park
KL	Kuala Lumpur
LAN	Local area network
LCD	Liquid crystal display
LED	Laser-emitting diode
LOC	Lead-on-chip
MFM	Magnetic force microscopy
MIGHT	Malaysian Industry Government Group for High Technology
MFN	Most favored nation
MIMOS	Malaysian Institute for Microelectronic Systems
MOS	Metal oxide semiconductor
MOSFET	Metal oxide semiconductor field effect transistor
MTC	Magnetic Technology Centre [Singapore]
MTDC	Malaysian Technology Development Cooperation
NIE	Newly industrialized economies
NTU	Nanyang technological university
NUS	National University of Singapore
OHQ	Operational regional headquarters (OHQ)
OECD	Organization for Economic Corporation and Development
NSTB	National Science and Technology Board [Singapore]
PAP	People's Action Party [Singapore]
PC	Personal computer
PCB	Printed circuit board
PCBA	Printed circuit board assembly
PCMCIA	Personal Computer Memory Card International Association
PDIP	Plastic dual in-line package
PDS	Part development system [Mentor Graphics]
PGA	Pin grid array
PLCC	Plastic leaded chip carrier
PTH	Plated through hole
QFP	Quad flat pack
RCU	Research and Consultancy Unit [UTM]
ROM	Read-only memory
RSE	Research scientists and engineers
SAM	Scanning acoustic microscopy
SBB	Stud bump bond

SCADA	Supervisory control and data acquisition system
SEM	Scanning electron microscopy
SIMM	Single in-line memory module
SIRIM	Standards and Industrial Research Institute of Malaysia
SMT	Surface mount technology
SOJ	Small outline J-leads
STM	Scanning tunneling microscopy
TAB	Tape-automated bonding
TBP	Total business planning program [Singapore]
TCM	Tunneling current microscopy
TCP	Tape carrier package
TI	Texas Instruments
TQFP	Thin quad flat pack
TSOP	Thin small outline package
UMNO	United Malay Nationalist Organization
USM	Universiti Science Malaysia
UTM	Universiti Teknology Malaysia
UTQFP	Ultra-thin quad flat pack
VLSI	Very large scale integrated circuit
WTO	World Trade Organization

Chapter 1

INTRODUCTION

Singapore and Malaysia are relatively unique among the countries of Southeast Asia. Both have good trade balances, are economically sound, enjoy good capital flow, are internationally competitive, have a diversified economic base that includes industry, are not heavily indebted to the International Monetary Fund or World Bank, and are expanding rapidly into the world market. And where Singapore is already an economic powerhouse of the region, Malaysia is well on its way to becoming one.

1.1. HISTORY

In the 19th century, the British saw the advantages of the port city of Singapore as a trading post in Southeast Asia. It provided a centralized location from which they could spread out into the region. It was also an area that provided them with the excellent transshipment facilities that they needed prior to making the long return voyage to England around the Cape of Good Hope and, subsequently, through the Suez Canal. Consequently, Singapore became a hub of commerce and a major outpost of the British Empire. In 1946, Singapore became a crown colony. It attained full internal self-government in 1959.

Malaya, as the area used to be named, also offered rich trading opportunities that attracted British occupation from the late 18th century through World War II. The Union of Malaya was formed in 1946 under British rule, combining the Federated Malay States, the Unfederated Malay States, and two of the Straits Settlements. It became the Federation of Malay in 1948 and attained full autonomy within the Commonwealth of Nations in 1957.

In 1963, Singapore joined the Federation of Malaysia, which then included Malaya, Singapore, Sarawak, and Sabah. This union only lasted two years; on August 3, 1965, Singapore withdrew, and Malaysia and Singapore separated into two independent countries. This separation was driven, at least in part, by conflict between the two major racial groups present in both countries, Chinese and Malay. The Chinese, who tended to be better educated and to control most of the capital in both countries, were the majority in Singapore and dominated its government but were a minority in Malaysia. The Malays, who were primarily involved in agricultural pursuits, were the majority in Malaysia and generally controlled the government there, but were a minority in Singapore. There were serious and recurring conflicts between the two groups, within each country, and also between the two countries.

This conflict manifested itself in 1969 when Malaysia was torn internally by rioting, primarily of Malays against the Chinese and Indian populations who owned a disproportionate amount of the country's wealth. The riots in Malaysia were the basis of a major turning point in Malaysian development. The army stepped in to help quell the riots, and emergency rule under Tunku Abdul Razak was established. Centralized rule by dictate continued for two years before some degree of relaxation occurred. From this point, however, Malaysia began its climb to economic power, as the army led or drove the country along the path of modernization. With a series of economic growth plans, the leadership under emergency rule diversified the country's economic base, constructed the necessary supportive infrastructure for development, mobilized the country's resources (particularly the population), and then turned the government back over to normal civilian authority.

The Republic of Singapore is a city-state with a parliamentary form of government in which a cabinet is drawn from the majority party in the legislature. From its inception, this party has always been the People's Action Party (PAP). Though the party and the culture of Singapore are dominated by the Chinese, who comprise 78% of the Singaporean population, Singapore officially proclaims itself a multiracial and multicultural society. Merit is the avowed basis of advancement.

Malaysia is a constitutional monarchy. The king heads the nation along with a bicameral legislature, consisting of a senate and a house of representatives. Members of the legislature come from the 13 states and two federal territories. The dominant political party has been and continues to be the United Malay Nationalist Organization (UMNO). This party, as is the culture, is dominated by Malays.

1.2. GEOGRAPHY

The Republic of Singapore is located in Southeast Asia off the southern tip of the Malay Peninsula between the South China Sea and the Indian Ocean. It consists of the main island of Singapore and 58 nearby small islands. The combined area of the country occupies slightly less than 250 square miles, with Singapore Island covering 93% of it. The island and the city of Singapore, essentially, are one.

Malaysia is separated into two regions by the South China Sea. The two regions are Peninsular Malaysia, on the southern part of the Malay Peninsula; and Sarawak and Sabah, located on the northern portion of the island of Borneo. The combined area of the two regions covers 127,317 square miles, an area slightly larger than the state of New Mexico. The capital and largest city is Kuala Lumpur.

Both Singapore and Malaysia enjoy easy access to the major sea lanes of Southeast Asia and a central location within the region. Construction of supportive infrastructures for economic development has optimized their geographic locations. Their major geographical drawback has been that they have few natural resources. Singapore, of course, is very small, and Malaysia's natural resources consist mainly of petroleum (in small amounts), timber, tin, and oil palm and rubber trees. Both countries, therefore, depend greatly on imports of raw materials for their industries.

1.3. DEMOGRAPHICS

Singapore's estimated population in 1994 was 2,900,000, yielding a population density of 11,755 people per square mile. Approximately 78% of the population are Chinese, 14% are Malay, and 8% are Indian, Pakistani, and Sri Lankan. The major religions are Buddhism, Daoism, Islam, Hinduism, and Christianity. Singapore's major cultural ethic, Confucianism, dominates the underlying system of thought and conception of the nature of relationships, specifically of the state to the people and vice-versa. There are four official languages in Singapore: English, Malay, Chinese, and Tamil. English is used by more than 25% of the adult population and is the principal language of business, government, and education.

Malaysia's estimated population in 1995 was 20,125,000, yielding a population density of 158 persons per square mile. Approximately 59% of the population are the indigenous Malays, who are predominately rural; 32% are Chinese, who are primarily urban dwellers; and 9% are a mixture of Indians and non-Malays of indigenous tribal groups. As in Singapore, the major religions of Malaysia are Buddhism, Daoism, Islam, Hinduism, and Christianity, but Islam is dominant. Although Malays dominate the government of Malaysia, the Chinese dominate commerce and banking. The official language is Bahasa Malaysia, but Chinese, Tamil, and English are spoken as well. English, again, is an important language of government, business, and education.

One of the major goals of Malaysia's development plans has been to weld the population together into a single unified nation instead of a merely geographic union of disparate racial and ethnic groups. In the book, *Vision 2020* [abdul Hamid 1995], which discusses from an official perspective Malaysia's goals through the year 2020, Prime Minister Mahathir Mohamad renews the pledge to create a unified populace who are all committed to the single goal of Malaysian development.

1.4. ECONOMY

Over the last 30 years, Singapore has experienced the most dramatic economic transformation of any country in East Asia. Its development stems from an excellent geographical location, enhanced by first-rate transportation facilities and from aggressively promoting the establishment of multinational electronics companies, which have aided Singapore in becoming an important competitor in the global electronics sector. Singapore's current strengths in the field of electronics are disk drives, semiconductors, surface mount technology, process control instrumentation, computers/communications/commerce electronics, and office automation.

During approximately the same period, Malaysia has built one of the healthiest economies among the developing nations of East Asia. In the process, it has used its agricultural base and earnings from offshore petroleum operations and from the timber and palm oil industries (and earlier, from rubber and tin production before these were supplanted as revenue staples) to support development of a primarily export-oriented economy. As this economic transformation began to occur, Malaysia has continued to develop its agricultural base but has increasingly emphasized industrialization. The country's current economic strength is due largely to its thriving export industry, which includes exports of electronic equipment and semiconductor devices. Malaysia is attempting to fortify its international standing by expanding its electronics industry through improving its educational facilities, placing greater emphasis on its research institutions, attracting multinational corporations, and updating the government's industrial policy. Just as the Republic of Korea, Taiwan, Singapore, and Hong Kong are dubbed the four "mini-dragons" of East Asia, Malaysia is clearly now a mini-dragon in the making.

Both Singapore and Malaysia are active members of the Association of Southeast Asian Nations (ASEAN) and the World Trade Organization (WTO). These memberships markedly contribute to their current and future export-import successes, giving them a competitive economic edge in regional trade over nonmember countries, including the larger industrialized countries.

The United States is the largest single source of capital for Singapore and one of the top three for the region, while Japan is another large source for both Singapore and Malaysia. Both countries have been major beneficiaries of Japanese investment, in both capital and in production facilities, and consider their trading relationship with Japan as vital.

1.5. EDUCATION

In both Singapore and Malaysia, education has been and continues to be an essential tool for development. The Confucian ethic in Singapore stresses the importance of education as paramount. Singapore's literacy rate is high, and education is popularly seen as a necessity for job acquisition and advancement. There is a close working relationship between the government and the National University of Singapore and Nanyang Technological University. Research and development (R&D) in Singapore is driven through the research institutes and centers, with the private sector playing a major role in identifying priorities and undertaking collaboration. As discussed in more detail in Chapter 6, the universities, institutes, and research centers not only provide the country with trained personnel and with research and design information, but they also regularly work on collaborative projects with the government.

In Malaysia, the breadth and depth of the emphasis on education within Malaysian society is not as great as in Singapore, but, as the *Vision 2020* plan discusses, it is seen by the government as crucial to further development. Approximately 78% of the adult population is literate (1990). In such universities as the Universiti Science Malaysia and Universiti Teknolgi, there is major governmental emphasis not only to train students to participate in the "real world" but also, as in Singapore, to establish collaborative research and development projects. In fact, Malaysian universities routinely allow companies to establish R&D groups on campus. In *Vision 2020*, the Prime Minister discusses the need for governmental investment in training Malaysians in the "work ethic" necessary to meet the employment needs of the country's development, as well as in the raw skills that various industries will need.

In each country there is a long-term commitment on the government's part to prepare the work force of tomorrow and the research and development infrastructure to meet the needs of their industries, and there is a high degree of collaboration between the government and the educational institutions. The governments have become major supporters of all facets of the educational systems.

1.6. CULTURE

Though dissimilar in many ways, Singapore and Malaysia have some very similar cultural perceptions underlying their growth strategies. Perhaps the first and most important of these shared perceptions concerns the role and nature of government in society. Both cultural ethics hold that government needs to be strong, centralized, and dominated by a single leader-group. (It is interesting to note that this shared perception is derived from two very different cultural and philosophical backgrounds.) With remarkable concurrence, the view in both cultures is that government serves to marshal resources, provide structure, and direct the course of society, for the good of all. There is room for some feedback from the people to the government, but when decisions have been made: it is the people's role to obey the directions of government and support it through concrete actions. This definition of the authoritative role of government and the subordinate, supportive nature of the people's relationship to it contributes to a singleness of societal purpose. Dissent tends to be comparatively infrequent and mild, because the population as a whole does not sanction it: questioning the authority of the government is not considered proper or respectful. This relational perception allows the government great latitude to plan the course its people will take.

Related to this is a second important shared cultural phenomenon: the intensity of the people's commitment to their countries' development programs. In Singapore, which enjoys one of the highest standards of living in Asia, this commitment by the majority of the general populace is easily observable in their diligent educational efforts, strong work ethic, and high savings rate. In Malaysia, popular commitment to rapid development and with it shared work and savings ethics may appear less strong, but the establishment of shared values that contribute to rapid economic growth is a government priority. As an example of its effort to foster popular commitment to development, the Government of Malaysia has established special preemployment schools to train new workers in these values prior to their starting employment with the developing industries.

A third major factor supporting development that Singapore and Malaysia share is the general commitment by the individual to the family (or clan or group), and to the society as a whole. The individual's responsibility to these outweighs personal desires. Thus, the people support the government and its direction because they perceive that the government's plans and actions are to the benefit of the society as a whole. This commitment of the individual to the social group also comes into play in the workplace; when employed, the people's allegiance to their employers is a deeply held matter of personal obligation.

1.7. SUMMARY

Several items are important to emphasize in this brief account of Singaporean and Malaysian development, particularly as they impact the business climates in these two countries:

1. The two countries are strongly driven by their respective cultures, Singapore by the dominant Chinese culture and Malaysia by the dominant Malay. Through the states' dominant political parties, PAP and UMNO, the ethics and perceptions of these groups dictate both their countries' economic development goals and the courses they will take to achieve those goals.

2. The two societies give complete authority to marshal the culture's resources directly to their governments, both of which have highly centralized decision-making systems.

3. Both governments are committed to intense short-term and long-term development, based primarily on exports. Their development scenario stresses economic diversification so that exploitation of natural resources, such as agriculture or oil, will account for an increasingly small percentage of gross national product (GNP) and industrial development will account for an increasingly large percentage of GNP.

4. In part due to the shortage of natural resources and the need to import most materials, the focus is on developing the value-added and services sectors. The electronics industry plays a major role in the development schemes of both Singapore and Malaysia.

5. If past performance is any indicator, both governments and societies will achieve their economic development goals.

Chapter 2

THE ELECTRONIC INDUSTRY'S ROLE IN SINGAPORE'S ECONOMY

Singapore's record of economic development in the years since its separation from Malaysia in 1965 has been perhaps the most remarkable in East Asia. Its success stems, to a great degree, from exploiting its excellent geographical location as a gateway to Asia, which is its appeal to foreign electronics companies that want to establish operations there.

2.1 BACKGROUND

In the 1960s, Singapore's economy depended almost entirely upon commerce and export trade. Since then, the country has developed a strong capability in manufacturing, services, telecommunications, utilities, and port services. With one of the best port and airline facilities in Asia in addition to its historically advantageous geographic location, it is a natural hub for trade and commerce.

During the late 1970s and early 1980s, multinational corporations (MNCs) tended to move to Singapore either for electronics assembly or for specialized labor-intensive work. However, Singapore did not make any strides from 1973 to 1990 in the value-added share of gross output, a measure of the level of technical development in manufacturing. In electronics, the ratio of value-added to gross output actually fell from 32.2% in 1978 to 27.7% in 1990. Manufacturing still depended on MNCs for technology, and very little research and development was undertaken on the island. Singapore's expansion in manufacturing output only occurred due to increased labor or capital, not to technical progress.

By the 1990s, almost every international electronics producer was represented in Singapore, and Singapore began to capitalize on technology transfer from the MNCs to local companies. In 1996, computers and computer peripherals, especially disk drives, became the key products of the nation's electronics industry with semiconductor following.

In order to further advance Singapore's technological ability and transition to more sophisticated value-added products, the government identified six development strategies that build upon the strong skills already derived from manufacturing, R&D, and marketing experience [de Scuza 1994]. The goal was to give Singapore the ability to respond to the changing trends in the electronics industry and its markets.

Strategy 1 - Develop industry clusters

The rapid growth of Singapore's electronics industry is to be sustained by three key industry groups: MNCs, technologically intensive local companies, and support industries. Representatives from each of these three groups form industry "clusters" around specific types of products or technologies. Singapore wants to strengthen the ties between these groups so that new sectors can emerge and new products evolve. Each cluster has gaps and opportunities to be identified and focused on.

The Economic Development Board (EDB) wants each member of a cluster to focus on codevelopment of new products. This will help local supporting companies, which have high proficiency in the production of parts and components, to become end-product manufacturers. The hope is that this initiative, focusing on strategic alliances, will become an engine for further growth.

Strategy 2 - Make Singapore a business hub

In order for MNCs to effectively participate in the rapid growth of Asian markets, they will either establish an operational regional headquarters (OHQ) and/or form partnerships with local companies on the island, based on the assumption that foreign companies will better understand the unique requirements of the region by being immersed in it. It will then follow, as customer needs are addressed, that mass customization and new products will emerge through focused R&D and process design based on awareness of the region's strengths and limitations.

This strategy primarily addresses growth in services by providing (1) a location for regional headquarters and business service provision; (2) logistics services; (3) a regionally centralized communications and media hub; and (4) support infrastructure such as education and healthcare.

In addition to opening regional markets, Singapore seeks to form partnerships with MNCs and share the risk of starting new businesses with them. As an example, Singapore's EDB shared the start-up costs and business risks with Texas Instruments (TI), Canon, and Hewlett-Packard (HP) to form the company TECH Semiconductor, which produces advance DRAM chips based on submicron CMOS technology. EDB and TI each own 26% of the venture, while Canon and HP each own a 24% share.

Singapore has used the OHQ program to attract over 50 MNCs to the region. The OHQ companies are expected to carry out R&D and high-value-added activities, such as technical and data communications services. They are also required to initiate incremental manufacturing projects. Motorola is one OHQ company that significantly upgraded its manufacturing and product design skills in Singapore.

Strategy 3 - Promote regionalization

One of Singapore's visions has been to build an external, regional economy that is strongly linked to, and augments, its domestic economy. The government plans to fortify its ties to other countries in the Pacific Rim for greater domestic economic growth. It also plans to build partnerships with regional companies, enhance the bond with developed countries and MNCs, and expedite government-initiated regional development plans. The ultimate goal is to position Singapore as the gateway to the region and to the world.

The EDB has already assisted companies in identifying international investment opportunities and harnessed the resources needed for expansion. Because of the EDB's support network, Singaporean companies can confidently penetrate emerging regional markets while allowing their domestic operations to expand into more value added activities.

Strategy 4 - Develop business councils

Singapore has demonstrated an enormous drive for tapping the vast business opportunities around the world, but the government realizes that it does not have the capacity to explore every opportunity; it needs international partners to profit from these opportunities. Consequently, Singapore has set-up joint business councils to promote and simplify cooperation. So far, France, Germany, and England have formed joint councils with Singapore.

Strategy 5 - Build second "S" corporations

Since Singapore has a booming consumer market, the demand for goods and services is expected to increase sharply. Most well-developed companies are at the top of the "S" curve, where it is difficult to pinch incremental sales. On the opposite end, many fresh companies, such as the new companies involved with electronics at the time these strategies were devised, are at or near the launch-point of the "S" curve, which positions them well for a technology push.

The secondary "S" curve companies are encouraged to grow into Singaporean MNCs by developing strategic plans, new products, and bigger markets. The government supports these companies by granting them "pioneer" and/or "post-pioneer" status, and by implementing programs, such as the Total Business Planning (TBP) program, that support rapid local expansion. The goal is to help reposition domestic and multinational companies in Singapore to develop new capabilities that can lead to a second "S" curve of growth.

Strategy 6 - Collective national marketing

Singapore wants to sell itself on collective attractiveness. This means that the government, corporate sector, and academia work together to satisfy common goals. The government invests heavily in training and tertiary institutions so that schools can graduate better skilled laborers and engineers. The government also teams up with foreign governments and leading MNCs to run technology training centers. Government leaders hope that the technology training centers will develop specialist manpower in the following areas: precision engineering, factory automation, integrated circuit (IC) design, and surface mount technology.

To further amplify the technological competitiveness of Singapore's electronics industry, the government has established a few national "centers of excellence" in applied research and development: the Institute of Microelectronics, the Center for Integrated Circuit Failure Analysis and Reliability, and GINTIC's Division of Manufacturing Technology.

Singapore with this concentrated effort has become one of the strongest and most sophisticated of East Asia's modern high-tech economies. With near total employment, one of the highest standards of living in Asia, a highly diversified economy, and a strong investment in R&D, it is well positioned to continue its outstanding economic performance.

2.2 STATUS OF THE INDUSTRY

Singapore's electronics industry both competes with and complements the electronics industries of East Asia's other newly industrialized economies, such as Hong Kong, Taiwan, and South Korea. Singapore's current strengths are in disk drives, semiconductors, surface mount technology, and process control instrumentation. World-class key players in each of these technologies has a presence in Singapore.

In 1993, Singapore manufactured 23.3 million disk drives, which accounted for more than 40% of the world output. Singapore's disk drive industry continues to prosper, as exhibited by the expansion of Seagate operations and the debut of IBM's disk drive facility there. According to Singapore's EDB, the disk drive and data storage device sectors will continue to be key segments of Singapore's electronics industry. Another focus is wafer fabrication. Of the world's top 12 semiconductor producers, 5 have established facilities in Singapore (NEC, Fujitsu, Texas Instruments, Matsushita, and SGS-Thomson), as have others, including National Semiconductor and AT&T.

The microelectronics industry in Singapore has production capabilities comparable to those of South Korea and Taiwan but owns a broader base of technologies and caters to more diverse regional and world markets. The 1993 *World Competitiveness Report* ranked Singapore as the world's most competitive newly industrializing economy for the fifth consecutive year. Singapore has also been ranked as the world's third best investment location by Business Environment Risk Intelligence (BERI), the United States-based risk assessment and consulting firm. Additionally, Singapore's infrastructure is outstanding. The facilities of its system of deep seaports and of its airport are among the best in Asia. Its rapid transit system of busses, trains, light rails, and roads provides easy and ready access to all of the major manufacturing, business, shopping, and residential areas of the main island. Its power and water supply system, shared with Malaysia and Indonesia, though increasingly expensive, is stable and secure. And its telecommunications system is completely digital and utilizes the most modern technology the country can buy. Singapore seeks to be a totally self-sufficient, self-sustaining society. Its leaders want the country to be a major regional hub and to attract value-added activities that will efficiently utilize its relatively small population.

Singapore is marshaling all facets of its society to pursue a course of even more intense development and diversification. With its very large resources of capital, highly trained and committed populace, and effective governmental-university-industry coalition, all committed to this goal, and major governmental programs to attract investment, technology-transfer, and business operations, Singapore is well on the way to achieving its goals. A colleague country, "Singapore's objective is to continuously improve the standard of living of its citizens. [It has] developed a comprehensive plan to diversify the economy into greater value-added products and services ... [and has] focused the entire country's organizational structure to do so, from Parliament to the Finance and Capital Market, from educational institutions to transport authority." [Kelly 1996]

2.3 ECONOMIC INDICATORS

In the three years 1992 to 1995, GDP grew 9.9%, 10.1%, and 8.5%, as a primary measurement of Singapore's economic growth. The government's goal is a 7% yearly growth in GDP while maintaining a low inflation rate of 2-4% and still recruiting significant investment (U.S. $2.4 billion in 1994).

The electronics industry plays a pivotal role in Singapore's economy. In 1993, the electronics industry accounted for 46% of the manufacturing sector's output and 34% of the labor force. The presence of MNCs in Singapore's electronics industry has enabled the country to develop a strong export-oriented manufacturing environment. MNCs have also helped local industries to diversify. Development of the electronics industry sector is likely to significantly impact Singapore's growth and development with respect to its neighbors in the Pacific Rim and the rest of the world.

2.4 MANUFACTURING IN SINGAPORE'S ECONOMY

Manufacturing, specifically in electronics, is key to Singapore's rapid growth rate. Manufacturing as a percentage of gross domestic product

(GDP) steadily increased from 1960 through 1991, where it peaked at 29.3% (Figure 2.1). In 1992, it dropped to 26.3%, and it has remained at approximately this point since then (1993-26.4%, 1994-26.1%, and 1995-26.7%).

In the period 1960 to 1995, Singapore's economy as a whole grew substantially, with the result that it can now be considered one of the most highly developed industrial, commercial, financial, and consumer economies in the world. Singapore's EDB illustrates in Figures 2.2a and 2.2b[1] how manufacturing is a key factor in the economy's growth.

In 1960, (Fig 2.2a) as 18% of the GDP of $1.5 billion, manufacturing accounted for S$378 million. The other main income areas were commerce, finance and business, and communications. After "others," the largest sector was commerce. In 1995 (Fig 2.2b), the largest sector was finance and business, 29% of GDP, accounting for S$34.5 billion, closely followed by manufacturing, which had grown to a proportion of 27% of a GDP of $79 billion and was valued at $22.9 billion. Commerce, having dropped to a share of 18% of the GDP, still contributed $15.3 billion to the economy.

Figure 2.3 shows how manufacturing has maintained a relatively steady growth rate, absorbing a larger and larger share of the workforce. This figure also illustrates how level of output and worker usage have constantly increased, despite the fact that population size, particularly after 1990, has remained relatively constant.

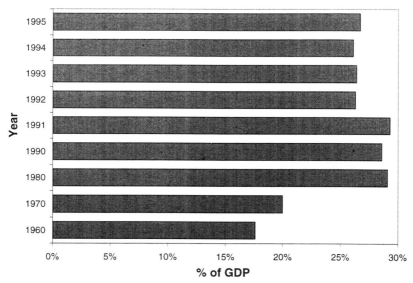

Figure 2.1. Manufacturing as a Percentage of GDP (from Manufacturing 2000: The Cluster Development Approach, Economic Development Board, Singapore, 1995).

---

[1] Throughout this book, where Singapore dollars are given, U.S.$1.00 = S$1.41.

MANUFACTURING IN SINGAPORE'S ECONOMY

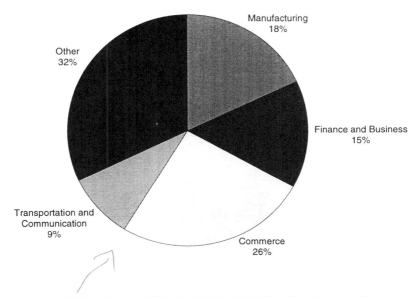

Figure 2.2a. 1960 breakdown of GDP, Total GDP = $1.5 billion (from Economic Development Board (EDB)).

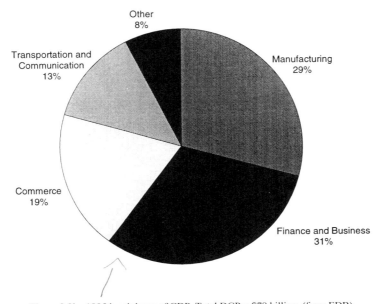

Figure 2.2b. 1995 breakdown of GDP, Total DGP = $79 billion (from EDB).

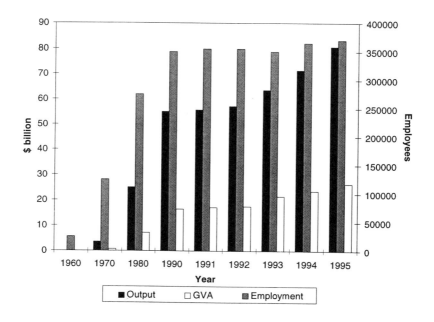

Figure 2.3. Growth of the manufacturing sector (from EDB).

Figure 2.4 clearly illustrates how productivity in the manufacturing sector has constantly increased, in terms of both output and gross value added (GVA) per worker. Driving this increase are education, technology, and increased value of products (through inflation, market changes, etc.).

2.5 ELECTRONICS IN SINGAPORE'S ECONOMY

Within the manufacturing sector as a whole, electronics has accounted for the largest share of GDP in Singapore's economy. As Figure 2.5 illustrates, in 1995, electronics accounted for 51.4% of total manufactured output, followed consecutively by chemicals, fabricated metal, telecom equipment, other machinery, electrical machinery, and others.

ELECTRONICS IN SINGAPORE'S ECONOMY

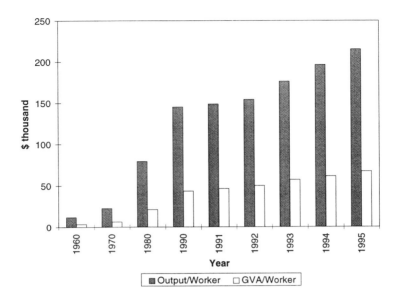

Figure 2.4. Productivity of the manufacturing sector (from EDB).

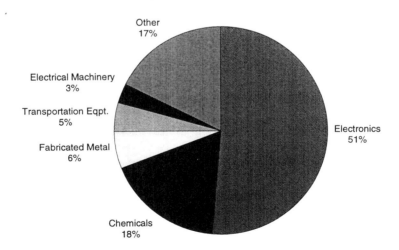

Figure 2.5. Output of Manufacturing Sectors (from EDB).

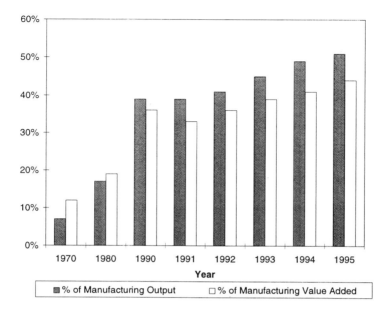

Figure 2.6. Dominance of electronics in manufacturing sector (from EDB).

Figure 2.6 shows how electronics has quickly come to represent a significant, then a major, share of all manufacturing output. In 1970, electronics was 7% of total manufacturing output and 12% of total manufactured value added. As is clear, particularly from 1990 on, electronics has occupied a larger and larger share of both total manufacturing output and total manufactured value added. By 1995, electronics accounted for 51% of total manufactured output and 44% of total manufactured value added.

Figure 2.7 also illustrates this point by comparing the rates of changes in output, GVA, and employment from 1994 to 1995. Compared to 1994, the output rate of change was 19% with 1995 output at $41.5 billion; the GVA rate of change was 20% with 1995 GVA at S$11.1 billion; and the rate of employment change was 4.2% with 1995 employment at 128,700. Figure 2.7 also clearly shows how productivity and GVA maintained a constantly higher rate of change than did employment.

With productivity and GVA rates consistently rising faster than employment (Figure 2.7), and with the current full employment in Singapore and increasing industrial growth, among other factors, there is a growing shortage of labor in Singapore, which is likely to get worse. As of 1995, Singapore already employed 450,000 Malaysians.

ELECTRONICS IN SINGAPORE'S ECONOMY

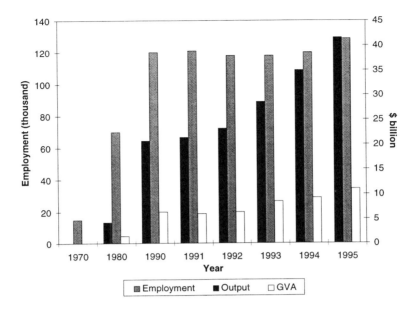

Figure 2.7. Growth of Electronics Sector (from EDB).

Figure 2.8 more clearly shows that the rate of output/worker (productivity) is rising much more quickly than the rate of GVA/worker in the electronics sector.

Figure 2.9 shows the relative strengths in 1995 of various subsectors of the electronics industry. The products that accounted for the largest share of production were data storage items (23%, $9.6 billion), closely followed by semiconductors (21%, $8.7 billion) and computers (20%, $8.3 billion), then followed by consumer electronics (12%, $5 billion).

As shown in Table 2.1, the semiconductor subsector has accounted for an increasingly large share of the industry's output: it experienced a 53% growth rate from 1994 to 1995. Interestingly, communications equipment jumped from a production share of 2.7% to 3.7%, an increase of 34%.

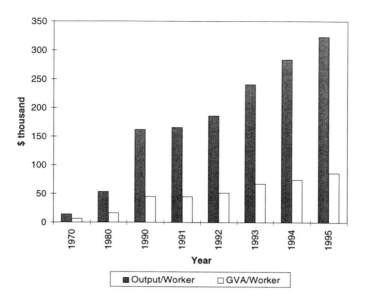

Figure 2.8. Productivity of electronics sector, 1970-1995 (from EDB).

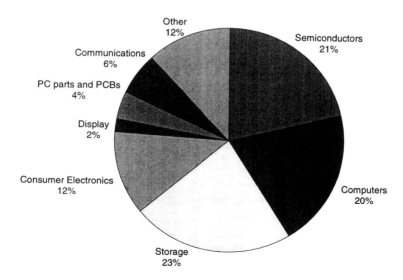

Figure 2.9. Breakdown of electronics output; 1995 electronics output total = $41.5 billion (from National Survey of R&D).

Table 2.1
Output of Singapore's Electronics Sector, 1994-1995

	1995 ($B)	1994 ($B)	% Growth 95/94
Data Storage	9.6	8.1	20
Semiconductors	8.7	5.7	53
Computers and MM	8.2	7.4	10
Consumer Electronics	5.1	5.3	-3
Communications	2.6	1.9	34
Office Automation	2.4	2.1	17
Passive Cpts and PCBs	1.5	1.4	10
Display Devices	0.9	0.7	21
Contract Mfg. & Others	2.4	2.4	0
Total	**41.4**	**35**	**19**

source: EDB

Singapore's current core capabilities in its electronics industry in fields such as digitization, wireless technology, miniaturization, automation, human interface technology, product intelligence, and product management are supported by a technological infrastructure composed of the following: the National Science and Technology Board, which spearheads R&D efforts with a strong focus on industries in Singapore; the Institute of Microelectronics; the GINTIC Institute of Manufacturing Technology; the Center for Wireless Communications; and the Data Storage Institute.

Manpower development, as part of Singapore's core electronics industry base, is developed through several programs: the Radio Frequency Engineer Development Program; the Analog IC Design program, jointly conducted by IME and the universities; specialized process technology courses in universities and polytechnical institutes for wafer fab, disk media, and IC packaging; conversion courses for Applied Science graduates; and INTECH grants for on-the-job training in product and process technologies. These efforts are supported by a $700 million Cluster Development Fund, the Innovation Development Scheme, the Local Industry Upgrading Program, neutral sites for MNCs to form joint ventures, and Partnerships for Regionalization.

A look at Singapore's R&D expenditures by sector underlines the above discussion (Table 2.2 and Figure 2.10). The private sector has always maintained a high R&D investment rate. This investment rate, since 1990, has escalated dramatically, as has investment in public research institutes. Prior to 1990, research in these institutes did not exist. Educational research has also been rather constant. Yet, as Figure 2.10 illustrates quite clearly, private sector R&D investment accounts for by far the largest degree of investment in the total economy. Table 2.3 and Figure 2.11 even more dramatically illustrate the nature of these relationships.

Table 2.2
Singapore R&D Expenditure by Sector (in $ million)

Year	Private Sector	Higher Education Sector	Government Sector	Public Research Institutes	Total
1978	18	6	3	*	27
1981	32	17	9	*	58
1984	76	50	27	*	153
1987	161	68	38	*	267
1990	221	86	71	31	409
1991	316	105	69	51	541
1992	413	111	75	79	678
1993	442	112	76	83	713
1994	526	128	101	84	839

- data not available prior to 1990;
- from National Survey of R&D in Singapore, 1994, National Science and Technology Board, Singapore Science Park, Singapore, 1995

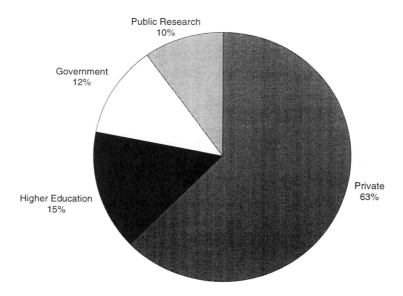

Figure 2.10. Proportion of electronics R&D expenditure, by sector, 1994 (from National Survey of R&D).

Table 2.3
Singapore R&D Expenditure by Strategic Focus, 1994 ($ million)

Type of Research	Private Sector	Higher Education Sector	Government Sector	Public Research Institutes	Total
Pure basic	4	12	2	17	35
Strategic basic	20	36		8	71
Applied	175	52	64	32	323
Experimental development	327	28	28	27	410
Total R&D expenditure	526	128	101	84	839

source: National Survey of R&D

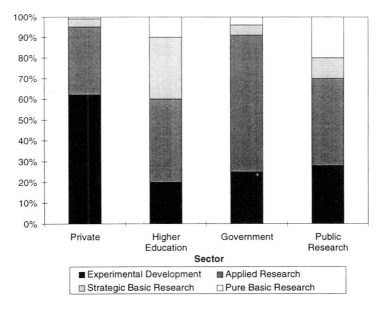

Figure 2.11. Breakdown of R&D Expenditure by Sector, 1994 (from National Survey of R&D).

Also, within the overall area of R&D research, Table 2.4 and Figure 2.12 show the breakdown of expenditures by area of emphasis. Within this breakdown, pure and strategic basic research together account for about 13% of the total R&D amount spent, while 53.9% is spent on engineering and 23.9% on computer and related sciences (Fig. 2.12); combined, these two areas account for 77.8% of all R&D expenditures.

Table 2.4
Singapore R&D Expenditure by Area of Research, 1994 ($ million)

Area of Research	Private Sector	Higher Education Sector	Government Sector	Public Research Institutes	Total
AGRICULTURAL SCIENCES	2.9	0.1	5.9	0.0	8.9
COMPUTER & RELATED SCIENCES	146.2	12.0	17.1	25.3	200.6
ENGINEERING	310.2	54.4	51.3	36.5	452.5
Aeronautical	0.3	0.0	0.4	0.0	0.7
Marine	1.9	0.0	0.5	0.0	2.4
Mechanical	98.8	15.0	4.5	0.0	118.3
Biomedical	2.7	3.4	0.0	0.0	6.1
Civil & Architecture	2.5	17.4	1.9	0.0	21.8
Electrical & Electronic	180.4	8.3	39.1	36.5	264.3
Material, Chemical Sciences	20.7	8.2	4.2	0.0	33.1
Metallurgy & Metal	2.9	2.1	0.8	0.0	5.8
BIOMEDICAL SCIENCES	0.9	24.9	10.6	0.0	36.4
NATURAL SCIENCES	51.0	36.1	4.8	21.9	113.8
Biological Sciences	12.3	13.9	2.7	21.9	50.8
Chemical Sciences	33.3	6.4	0.8	0.0	40.5
Earth & Related Environmental Sciences	1.0	1.6	1.3	0.0	3.9
Physical Sciences & Math	4.4	14.2	0.0	0.0	18.6
OTHER AREAS	14.6	0.7	11.9	0.0	27.2
TOTAL R&D EXPENDITURES	525.8	128.2	101.7	83.7	839.4

Source: National Survey of R&D

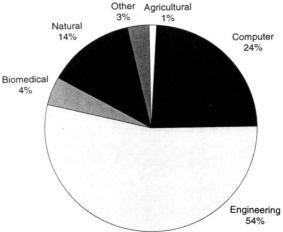

Figure 2.12. Proportion of R&D expenditure, by area of research, 1994 (National Survey of R&D).

Table 2.5, which examines the sources of funding for this research, emphasizes that the research dollars come primarily from private industry and the Singaporean government. Internal source funds and governmental funds account for over 90% of all funds invested.

Table 2.6 and Figure 2.13 show the levels of organizational support by country. It is significant that of all countries supporting Singaporean R&D, only 53.9 % are Singaporean. Figure 2.14 shows that the bulk of all R&D moneys are spent in various electronics fields: electronic equipment and appliances account for 48.7% of the total private sector R&D expenditure by industry group.

Finally, Table 2.7 shows foreign companies doing business in Singapore by majority ownership and industry group. This illustrates several points necessary to understanding Singaporean development due to capital and technology flows. Firstly, the major foreign investors in Singapore's manufacturing sector, with emphasis on the subsector of electronics, are the United States and Japan. Secondly, within the electronics subsector, their major area of investment has been in equipment and components. And thirdly, the next largest area of investment for both the United States and Japan been the service area. The two countries combined account for 59% of all foreign companies in Singapore that have majority ownership.

In examining these statistical relationships, several items become clear:

1. Within the overall Singaporean economy, manufacturing dominates the other sectors in degree of investment, growth, and rate of positive change.

2. Within manufacturing, the electronics sector dominates all of the other subsectors in investment, growth, and rate of positive change. Therefore, the electronics industry, by individual sector, has been and continues to be the major growth sector in the Singaporean economy.

3. R&D is a major point of concern among all sectors of Singapore's economic development support structure, including government and universities. Within this support structure, basic research receives a significant amount of funding, primarily from government and university sources, but the majority of R&D money goes to manufacturing and electronics.

4. There is very heavy foreign direct investment in the Singaporean economy, primarily coming from the United States (the largest foreign investor) and Japan. Their areas of investment, particularly through majority ownership, are identical for both countries (manufacturing — primarily in electronics — and services).

Table 2.5
Singapore's Source of R&D Funds, 1994 (in $ millions)

Sector	Internal Sources		External Sources					Total
	Own Funds	Other Internal Funds	Other Companies, Locally Based	Other Companies, Foreign Based	Local Tertiary Institutions	Singapore Government	Other Governments	
Private Sector	449.3	42.7	1.1	20.3	0	12.3	0.1	525.8
Higher Education Sector	1.0	0.0	0.7	0	0	126.4	0.1	128.2
Government Sector	23.4	1.9	0.0	0	0.0	76.4	0	101.7
Public Research Institutes	1.8	0.0	5.6	0	0.0	75.7	0.6	83.7
TOTAL	475.5	44.6	7.4	20.3	0.0	290.8	0.8	839.4

Source: National Survey of R&D

Table 2.6
Number of Organizations Involved in R&D Cooperation (1994)

Nature of Cooperating Institutions	USA	Japan	Europe	ASEAN (Excl. S'pore)	Others	S'pore	Total
Parent Company	76	42	54	3	8	-	183
Tertiary Institutions	13	5	16	0	21	117	172
Public Research Institutes	1	3	8	0	7	97	116
Buyer Organizations	7	4	6	8	19	127	171
Suppliers	29	12	30	2	11	127	211
Private R&D Organizations	3	5	6	1	8	37	60
Other Private Organizations	20	4	15	6	16	173	234
Other Associate Companies of the Group	33	15	54	37	61	105	305
All Institutions	182	90	189	57	151	783	1452

Source: National Survey of R&D

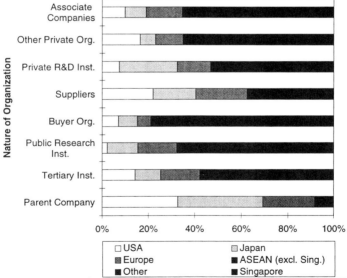

Figure 2.13. Percentage of cooperative R&D arrangements (from National Survey of R&D).

Table 2.7
Foreign Companies in Singapore by Majority Ownership and Industry Group

INDUSTRY GROUP	USA	Canada	United Kingdom	France	Germany	Italy	Japan	China	South Korea	Taiwan	ASEAN (Excl. S'pore)	Others	TOTAL
AGRICULTURE & FISHING	0	0	0	0	0	0	0	0	0	0	0	0	0
MANUFACTURING	49	1	9	6	6	3	32	1	0	0	2	31	140
Electrical / Electronic	27	0	2	4	4	1	12	0	0	0	0	12	62
Electrical Machinery	6	0	2	2	2	0	2	0	0	0	0	3	17
Electronic Equipment & Components	21	0	0	2	2	1	10	0	0	0	0	9	45
Chemicals / Petrochemicals	13	0	7	2	0	0	7	1	0	0	0	11	41
Chemicals & Chemical Products	11	0	6	2	0	0	6	1	0	0	0	10	36
Pharmaceuticals	1	0	1	0	0	0	1	0	0	0	0	1	4
Petroleum Products & Refining	1	0	0	0	0	0	0	0	0	0	0	0	1
Chemical-Linked	2	0	0	0	0	0	5	0	0	0	1	0	8
Food Products, Beverages, & Tobacco	0	0	0	0	0	0	4	0	0	0	1	0	5
Rubber & Plastic Products	2	0	0	0	0	0	1	0	0	0	0	0	3
Transport Equipment	1	0	0	0	0	1	0	0	0	0	0	1	2
Basic / Fabricated Metals	1	0	0	0	1	1	1	0	0	0	0	0	4
Machinery / Instrumentation Equipment	5	1	0	0	1	1	7	0	0	0	1	5	21
Other Manufacturing Industries	0	0	0	0	0	0	0	0	0	0	0	2	2
CONSTRUCTION	0	0	0	0	0	0	0	0	0	0	0	1	1
SERVICES	26	2	6	4	3	0	16	0	0	0	1	9	67
Information Technology	5	2	3	2	1	0	2	0	0	0	1	1	17
Other Services	21	0	3	2	2	0	14	0	0	0	0	8	50
ALL INDUSTRY GROUPS	75	3	15	10	9	3	48	1	0	0	3	41	208

Source: National Survey of R&D

ROLE OF TRADE

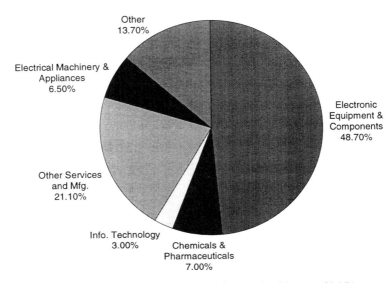

Figure 2.14. Private R&D expenditures (from National Survey of R&D).

If all trends continue as they are currently, the Singaporean electronics sector will be the primary source of Singaporean development in the future, both as a percentage of total development and in total dollar amounts. Two contributing factors are that capital investment in this sector from all sources continues to grow in both total amount and percentage of investment, and that focused R&D programs continue to provide a solid basis for the continuation of this general trend.

2.6 ROLE OF TRADE

During the 1970s, as manufacturing became the economy's leading sector, Singapore moved to export-oriented industrialization to grow away from its earlier dependence upon a staple port economy. In doing so, Singapore continued to take advantage of trade as an engine for growth by exporting manufactured goods to the West. Electronics assumed a larger and larger share of this role. The country operated one of the world's most successful less-developed economies by taking advantage of the new production function offered by trade.

A major factor in Singapore's stratagem for achieving growth through exports is its membership in various multilateral trading organizations. The General Agreement on Tariffs and Trades (GATT, now the WTO), ASEAN, etc., have all provided the country with significant trading advantages, such as the General System of Preferences (GSP), which the country's leadership has been quick to turn to advantage.

After 1980, the relative importance of manufacturing declined slightly, but the electronics industry maintained its rapid growth. Exports of office machines and telecommunications equipment grew at an annual average rate of 22%. By the end of the decade, Singapore was the world's largest exporter of Winchester hard disk drives.

Singapore is rapidly growing into a world-class electronics manufacturing center. Electronics account for the majority of the country's total manufacturing output as well as of its exports. Exports of locally-made electronic products and components reached a value of $25.3 billion in 1993. Singapore was the United States' ninth-largest export market in 1994 (US$15 billion), and the United States has remained Singapore's largest market for its electronic products, accounting in 1993 for 38% of its total domestic exports. In 1993, Singapore's other significant markets were Malaysia (8.1%), Germany (7.1%), Japan (6.6%), the Netherlands (5.3%), Great Britain (5.2%), and Hong Kong (4.8%).

Domestic electronics exports include industrial electronics, consumer electronics, and electronic components. Industrial electronics account for more than half of Singapore's total electronics exports. The increased demand for Winchester hard disk drives and computers contributed in 1993 to the growth of industrial electronics exports by 22% to reach $15.2 billion.

Exports of domestic disk drives grew by 6% to reach $6.7 billion in 1993, maintaining Singapore's position as the world's leading manufacturer of disk drives. Singaporean disk drive companies are automating their plants to be more cost-effective, and they are also introducing new products with higher memory capacity, such as 1.8-in. drives, to meet market demands. The coordinated support these companies have received from government, industry, etc., is the primary source of their continued dramatic expansion.

Export growth of computers increased by 47% to $6.4 billion in 1993. The computer companies are now involved in more product development activities. Singapore also has gained recognition for producing multimedia products, supplying more than 80% of the world demand. Companies there have successfully introduced many new products, such as notebook computers and sound cards, into the European and U.S. markets.

Despite the global economic slowdown in 1993, domestic exports of consumer electronics reached $4 billion in 1994. A number of companies rationalized their production by redistributing some of their labor-intensive production to other locations in the city-state. Their plants in Singapore now concentrate on higher-end equipment and components, such as optical pickups for CD players and magnetic cylinders for VCRs.

Domestic exports of electronic components grew by 24.2% in 1993, to $6.3 billion, although the growth in 1992 was only 7.4%. The semiconductor industry was the largest exporting sector in 1993, accounting in that year for 64% of total exports of electronic components. Since then, the semiconductor industry has continued to move towards greater

ROLE OF TRADE

integration of operations by introducing product and process development capabilities to strengthen assembly and test operations.

Singapore's economy, highly dependent on exports, is still on the fast track — and electronics dominates that course. Singapore is a prime example to its immediate neighbor, Malaysia, that small size and lack of natural resources are not insurmountable barriers to growth and development. Singapore's success also seems to indicate that an entrepreneurial society can, provided the citizens are supportive of the government's leadership, thrive within and in fact be tremendously aided by an essentially authoritarian political structure, to the benefit of the society as a whole. With one of the highest standards of living and per capita income in Southeast Asia, the country is proof that growth does not have to be at the cost of any major segments of society.

Chapter 3

THE ELECTRONIC INDUSTRY'S ROLE IN MALAYSIA'S ECONOMY

Pursuing a very pragmatic and focused course of development, Malaysia since its 1969-71 period of emergency rule has built one of the most robust and diverse economies among East Asia's developing nations. Its current strength is due largely to the country's thriving export industry, which includes electronic equipment and semiconductor devices. Growth in this sector has been supported by marketing its natural resources and by attracting large investments of capital. Malaysia aims to fortify its export posture by expanding its electronics industry, improving its educational and research facilities, attracting multinational corporations, and updating its national industrial policy. Rapidly and continually improving its strength and sophistication, Malaysia's economy is on the move.

3.1 BACKGROUND

With a free market economy and a relatively stable gross domestic product (GDP) growth rate of 8-9% per year (9.6% in 1995), it is clear that the Malaysian government's economic plans have been successful over a long period of time. Within the series of 3-5 year plans it inaugurated in 1969, the main areas of growth and development have been agriculture, tourism, and manufacturing. The country has become one of the world's leading exporters of semiconductors, air conditioners, rubber gloves, and consumer electronics.

The government of Malaysia has created a solid infrastructure that is capital investment friendly. It offers monetary exchange control regulations that allow the easy movement of funds, and it offers attractive tax incentives that include partial exemption from tax payments through designation of "pioneer status," investment tax allowances and abatements for exports, and deductions for R&D training. There is, however, a requirement for significant ownership participation by indigenous peoples before corporations can be listed on the stock market.

Malaysia enjoys a unique global trade status: despite its fairly sophisticated technological and developmental levels, it still falls under the WTO's Generalized System of Preferences (GSP). GSP status currently allows special negotiation rights in multilateral trade negotiation rounds and release from some of the article requirements on reciprocity and MFN (Most Favored Nation) trade provisions. Malaysia has the infrastructure, technological expertise, and manufacturing diversity to meet customer quality requirements sufficiently to pass the content hurdles contained in the agreement's articles.

Supported by its high level of government support, the Malaysian electronics industry has matured at a rapid rate over the last 25 years. Starting out with an investment and production base consisting of, and dependent on, multinational corporations (MNCs). Malaysia's advantages at the time were low-cost labor, stable government, capital inflow (from both governmental and nongovernmental sources), and a strong infrastructure. In a generation, the electronics sector has grown to employ over 200,000 people.

The electronics industry has recently taken aggressive strides to further accelerate its growth rates. From 1985 to 1990, the compound annual growth rate and output of its electronics industry was 23.2%; from 1989 to 1990, it was 40.8%. The output of value-added industrial sections almost tripled from 1985 to 1990, with a concurrent employment increase of over 160%. Figure 3.1 illustrates some of these trends for the period 1978-1990.

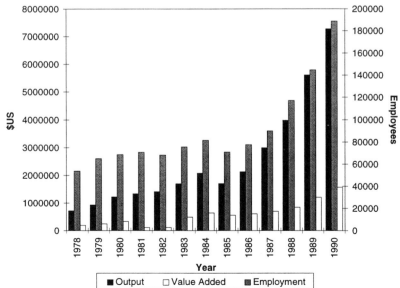

Figure 3.1. Growth of Malaysian electronics industry (From Shariffadeen, T.M.A., "Malaysian Electronics Industry - Transition and Transformation," IMS 93 keynote address, August 1993.)

STATUS OF THE INDUSTRY

According to Tengku Mohd Azzman Shariffadeen, Director General of the Malaysian Institute of Microelectronic Systems (MIMOS), one downside of Malaysia's rapid growth is that the value-added per person did not increase significantly from 1984 to 1990. In 1989 and 1990, the value-added per person was $8,300, compared to $7,800 in 1990. Total value-added has only been increased by raising the number of workers. This is not considered to be a long-term, cost-effective, or real solution; improving production functions, such as raising levels of automation, and increasing the level of production sophistication are understood to be the only long-term way to bring about a real increase in value-added cost effectiveness.

In the mid-1980s, the extreme dominance of electronic components, mainly semiconductors, and the dependence on this narrow subsector, was a cause of concern within the electronics industry as a whole. As a result, the Industrial Master Plan (IMP) was introduced with the goal of "balancing out" the industry. Table 3.1 shows the planned proportions of components, consumer electronics, and industrial electronics within the sector.

Malaysia reached and surpassed its 1995 number goals five years ahead of schedule. Happily, one target that the IMP grossly underestimated was industry employment. The target employment for 1995 was 149,420; instead, the industry created employment for 188,659 people, which is significantly above the amount targeted only five years earlier.

3.2 STATUS OF THE INDUSTRY

Currently the Malaysian Technology Development Corporation (MTDC) provides seed capital to companies interested in investing in Malaysia. The MTDC also supports Malaysian companies that are competing in high-technology areas. Specifically, the organization has four objectives:

Table 3.1. Malaysia's Electronics Industry Structure (Output)

Year	Electronic Components (%)	Consumer Electronics (%)	Industrial Electronics (%)
1986	81.5	12.3	6.2
1987	76.4	15.7	7.9
1988	71.3	18.0	10.7
1989	64.2	22.6	13.2
1990	57.6	23.2	19.2
IMP Target 1995	**61.0**	**24.0**	**15.0**

source: IMS 93 keynote address

1. develop local technology
2. aid in transferring to industry the results of the research and development performed at universities and research institutions
3. gain access to foreign technology that it can then transfer to local companies
4. develop venture capital for Malaysian industries

In the government's vision for the future, Malaysia will become a fully industrialized and developed nation by the year 2020. In order to get a good start on reaching this goal, the government has implemented the Vision 2020 plan, which targets the electronics industry as being pivotal for industrialization and development. Its strategies are to

- pursue export-led growth and free market strategies
- increase development of technology-intensive industries
- increase development of high-tech and brain-intensive industrialization and pursue an accelerated industrialization drive

In these strategies, the technology field generally recognized as having had the greatest impact on Malaysia's economy is electronics, including a wide range of specialized areas, such as design and manufacture of components, design and assembly of systems, and a blossoming range of increasingly sophisticated applications for industry, defense, and consumer products. Microelectronics and digital technology are two of the key areas in this field. Malaysia has acquired a high degree of familiarity with the electronics sector because of the manufacturing activities of its resident MNCs, but it is still working to generate domestic competence and innovative capability in this area.

The Industrial Technology Development Plan gives a summary of action profiles that Malaysia must take to obtain excellence in the electronics field:

- develop technological competence in semiconductor technology, microprocessor engineering, radio frequency engineering, surface mount technology, and digital signal processors
- enhance local design and development support for application specific integrated circuits (ASICs)
- promote development and design of high-value-added electronic equipment, such as printers, disk drives, facsimile machines, and multimedia products
- promote application of microelectronic technology in process automation services

It is arguable that the most significant part of the government's activities to promote economic growth is its carefully orchestrated and highly successful campaign to reeducate, unite, and modernize the Malaysian people, persuading them that this marshaling of resources and concentrated development effort is to the benefit of all in the society.

In the future, Malaysia's economic growth will have to rely as heavily on the private sector as it has on the public sector in the past. Although the government will maintain its leadership by providing a regulatory framework for rapid economic and social development, it wants to downsize its role in economic production and business. The government is also installing a support infrastructure that will ensure rapid, realistic, focused, and market-driven development of its electronics capabilities. As the Prime Minister stated in *Vision 2020* [abdul Hamid 1995],

> We have to accept half the technology and build on it and perhaps in time we will be ahead of them, as the Japanese are now in main areas ahead of the people who supplied them the original technology... [We will] double our per capita income, our GNP every ten years, and it has been calculated that if we achieve an average growth of 7% per year then we will be able to achieve an 800% growth over a period of thirty years.

Considering the country's current performance as well as its performance over the last three decades, the Malaysian government seems to have proven that it is capable of mobilizing the necessary resources and that it has the necessary commitment to perform this developmental feat. Vision 2020 lays down the basic components of a growth philosophy for developing nations: reeducation of worker's values to support greater productivity; training of personnel in high-tech skills as a value-added measure; garnering of increasing amounts of state and private capital; and committing the energies of every sector of society towards a development goal that increasingly emphasizes high technology.

Multinational corporations are increasingly taking advantage of Malaysia's pro-growth, pro-high-tech climate by transferring state-of-the-art technology and manufacturing into the country, particularly when the government is, as in some cases, offering 100% financing of ventures that can further enhance technology transference. For these MNCs, Malaysia offers the best of all possible worlds: a willing government that provides significant financial incentives; a good infrastructure and geographic location; and cultural as well as financial and political support for their efforts. They enjoy the maximum possible degree of government support with the minimum degree of interference. Malaysia also offers optimum local resources with which MNCs can pursue their individual strategies, with the guarantee of government action to help provide these resources and to continue to develop them for usage. Malaysia's enticements to industry far exceed what most of the rest of the world offers, including the major industrial powers.

3.3 ECONOMIC INDICATORS

In 1992, the gross national product (GNP) was $51.9 billion, and per capita income was $2,790, one of the largest in Asia. A labor usage breakdown shows that agriculture and fishing accounted for 26.8% of GNP, manufacturing accounted for 20.1%, commerce and services accounted for 23.5%, government and public authorities accounted for 12.5%, and construction accounted for 6.7%.

A breakdown of Malaysia's economic sectors reveals the planned expansion of its industrial and manufacturing sectors. Table 3.2 shows the planned changes over time in the relative weights of the main economic sectors. In 1990, the sectors had the following percentages: primary production, 28%; industry, 30.2%; manufacturing, 27%; and services, 41.8%. The goal for 2000 is that primary products will account for only 18.0%, industry for 38.7%, manufacturing for 37.2%, and services for 43.3%. Further out, the desired relationship by 2020 is that agriculture account for 7.6%, industry for 42.8%, manufacturing for 40.0%, and services for 49.6%.

Table 3.2
Malaysia's Structure of Production Compared with the World

Sector	Primary	Industry*	Manu-facturing	Services
Averages for				
Low-income economies†	33	28	14	39
Middle income economies	19	38	24	43
Developed Economies	3	34	20	63
Industrial Countries				
South Korea	10	44	26	46
Singapore	0	37	26	63
United Kingdom	2	37	20	61
United States	2	29	17	69
Germany	2	37	32	61
Japan	3	41	30	56
Malaysia				
1990	28.0	30.2	27.0	41.8
2000	18.0	38.7	37.2	43.3
2020	7.6	42.8	40.0	49.6
Post Industrial				
2030	5.4	36.0	35.0	58.6

Source: Abdul Hamid, Ahmad Serjii, 1995 (*Vision 2020*) (* Industry comprises manufacturing, construction, and utilities); († Excluding China and India)

Comparing Table 3.2 to worldwide growth trends, several important points should be stressed: (1) Of the four major economic sectors, industry and manufacturing are scheduled to have the largest growth in Malaysia in the next 25 years, although services, already a high percentage, will capture a greater total share. Worldwide, however, according to current World Bank, IMF, and other data, service industry growth as a percentage of GNP-GDP is increasing at a geometric rate, while growth in manufacturing and products industries is increasing at a linear rate and therefore is accounting for an increasingly small percentage of GNP-GDP. (2) If the "post-industrial" world emerges as anticipated over the next several decades, the manufacturing sector in the larger industrial economies will hold a smaller and smaller percentage of overall production, and services will capture a larger and larger percentage. However, in Malaysia, even while the services area grows, industry and manufacturing will continue to grow dramatically. (3) The role of agriculture in Malaysia will diminish dramatically, and agriculture's market share will approach that of the developed economies.

3.4 ELECTRONICS IN MALAYSIA'S ECONOMY

In 1990, electronics contributed 24% to GDP in Malaysia and 39% to the manufacturing sector's earnings. According to the Vision 2020 plan, as the electronics industry becomes stronger, electronics will capture an increasingly large share of both GDP and the manufacturing sector. This should drive up both per capita income and service industry production. Currently, the electronics industry growth rate is about triple that of the GDP, although Shariffadeen states that in the future, the electronics industry growth rate should become closer to the GDP growth rate. This indicates that Malaysia's reliance on electronics to elevate manufacturing and exports earnings is expected to decrease somewhat.

This does not predict a smaller electronics industry; just the contrary. The electronics industry will continue to grow at a very significant rate, as will the entire economy (at least, as expected in the Vision 2020 plan). It will just grow at a slower rate than it has been growing.

Referring back to Table 3.1, within the electronics industry itself, consumer electronics has steadily increased in proportion to electronic components and industrial electronics, so that in 1990 consumer electronics products were 23.2% of total electronics production in Malaysia, close to the IMP target of 24% and double the 1986 share of 12.3%. This appears to a be a relatively stable trend.

Other growth trends in Malaysia's electronics industry include slowing of the growth rate in semiconductor production, which was only about 3% from 1991-1992. Previously, semiconductors were the fastest-growing subsector. The fastest-growing subsectors in 1991-92 were audio electronics (45.5% growth rate), video electronics (15.4% growth rate), electronic data processing (EDP), and office equipment (59% growth rate 1991-92, 109.7%

growth rate 1990-91.) The outputs of these three subsectors, as a percentage of the total electronics market, were semiconductors at 39% ($ 5.4 billion), audio and video at 27% ($ 3.7 billion), and EDP and office equipment at 17% ($ 2.4 billion).

For the period 1988-89, the increase in electronics industry output was 40.8% (Table 3.3 and Figure 3.1). Output increased an additional 29.6% from 1989 to 1990. In the five-year period 1985-1990, the compound annual growth rate was 29.6%. From 1988 to 1989, value-added and employment increased 86% and 61%, respectively. These figures indicated a strong and healthy industry with some sectors being perhaps destabilizing as they increased sector dependence, since they were growing more rapidly than other parts of the sector. The IMP plans took this into account and manipulated sector growth to increase balanced diversification and overall stability of the electronics sector as well as of its individual subsectors.

In the three decades of development in Malaysia since 1969, the economy has continued to develop as a whole at a relatively constant rate of 8-9%. Economic statistics prove that the 3-5-year development plans of Malaysia's military, then civilian, governments have created a diversified, highly industrialized, high-tech, stable, and expanding economic base. The electronics industry is not as crucial a part of the overall economy as it once was, but it still enjoys a major role and is growing. A recent decrease in IMP emphasis on electronics is due at least in part to the sector's comparative strength, in order to promote diversification and to reduce dependency on any individual sector, as well as due to changes in the international marketplace.

3.5 ROLE OF TRADE

With a population of 20 million, Malaysia has an internal market too small to sustain growth, so it has placed great emphasis on cultivating export markets. Malaysia has become one of the strongest exporters in the world, with exports totaling $41 billion in 1993. Malaysia is now ranked the 19th largest trading nation in the world. Its rank climbs to third if the countries in the Organization for Economic Cooperation and Development (OECD) are excluded. The country consistently ranks in the top ten in trade balance with the United States. Manufacturing made up 31.5% of GDP in 1993, while export earnings from manufactured goods totaled 77.5% of total exports.

Also in 1993, Malaysia contributed 27.7% of total trade among the nations of the Association of Southeast Asian Nations (ASEAN), whose members are Indonesia, Malaysia, the Philippines, Singapore, Thailand,

Table 3.3
Malaysia's Electronics Industry: Output, Value Added, and Employment (1978 = 100)

Year	Gross Value of Output (1,000)	Value Added (1,000)	Employment (Persons)
1978	743,868	188,027	53,597
1979	964,818	245,266	65,013
1980	1,268,926	333,333	68,653
1981	1,387,514	113,254	70,683
1982	1,470,531	111,591	68,209
1983	1,767,772	485,889	75,698
1984	2,160,988	638,776	81,627
1985	1,771,267	555,383	71,924
1986	2,213,840	602,651	77,547
1987	3,109,072	692,998	89,994
1988	4,140,840	842,694	117,104
1989	5,829,998	1205,349	144,848
1990	7,557,146	1,569,371	188,659

Source: IMS 93 keynote address

Brunei Darussalam, and Vietnam. Trade with ASEAN countries in 1993 constituted 27.9% of total exports and 19.8% of total imports. This underscores Malaysia's desire to utilize multilateral trading organizations, such as ASEAN and the GATT-WTO, to enhance its competitive edge within the international trading system. It is both a leader in ASEAN and a leader of Third World countries in the GATT-WTO.

It is Malaysia's intention to become ASEAN's center of capital goods production (machinery and equipment). This goal can only be achieved if high-technology and R&D become the focus of Malaysia's industrial strategy. With the publication and pursuit of *Vision 2020*, this seems to be a definite part of the Malaysian government's strategy.

In order for Malaysian trade to remain competitive, productivities and efficiencies must be improved while reducing cost. This must be accomplished through technological upgrading by acquisition or absorption of new technology, innovation, and R&D. With its investment attraction plans, such as tax incentives, financing of industry up to 100% of cost if technology transfer is involved, infrastructure construction, and preparatory training of workers, Malaysia's government has shown itself to be serious about achieving these goals, as they are discussed in the Vision 2020 plan.

Although the commodities portion of Malaysia's materials market is easily swayed by shifts in the world economy, an effect endemic in commodities markets, Malaysia still places heavy emphasis on it. The commodities market, though decreasing as a percentage of exports, has been a major source of capital for development of the manufacturing sector, to include electronics. The country's chief commodities exports are natural rubber, palm oil, tin, timber, and petroleum.

Textiles and electronics account for most of the manufactured portion of Malaysia's exports. Electronics exports account for about one-third of total export dollars. It appears that the electronics export growth percentage is slowing down: it was 31.2% from 1990 to 1991 and 13.8% from 1991 to 1992, compared to all exports. By 1993, the government was not yet certain if this was a trend. It is arguable that this was a percentage of overall growth in exports: so the sector theoretically could be growing, but decreasing as a portion of the overall amount of exports.

In 1992, the three largest sectors of electronics exports were (1) semiconductors, (2) audio and video (AV) equipment, and (3) EDP and office equipment. The second two sectors, especially EDP and office equipment, were experiencing phenomenal growth in the early 1990s. The semiconductor sector, despite reduced growth rates, was continuing to hold the number one position for electronics exports. Malaysia is the world's number three producer of and number one exporter of semiconductor chips. From the early 1970s to about 1990, the semiconductor devices sector grew by approximately 20% annually. From 1991-1992, growth was only 3%, perhaps indicating that the semiconductor industry was stabilizing. Part of this slowing of growth rate is due to the government, in its IMP, trying to reduce the economy's too-heavy dependency on any one sector.

Additionally, through government efforts eleven Free Trade Zones (FTZs) have been created, which have been strategically located throughout the country. Within these zones, companies are subject to minimal customs formalities and are exempted from import duties on raw materials, machinery, and component parts used in manufacturing. These FTZs particularly cater to electronic industries; six of the eleven zones contain major electronics operations.

In summation, Malaysia's trade position is expanding, particularly in industrial production and services. Considering its starting point in 1969, Malaysia's current trade status is remarkable, particularly when compared to other Third World countries. The country's economic achievements to date lends credence to the statements in *Vision 2020* that Malaysia will continue to achieve rapid growth in the next few decades. The country has developed a stable and solid economic base from which to launch this effort: infrastructure has been developed and maintained; through the coordinated actions of the government, university and educational institutions, and private enterprise, worker training (in work ethic as well as job skills) has become a major effort; the government has gone to great lengths to attract capital and industry, with great success, as illustrated by the high level of Japanese investment; and the government's IMP plans have constructed a diversified and balanced economic base, spanning raw materials such as petroleum to very high-tech industries. Malaysia is working very hard to achieve the status of a modern industrial state.

Chapter 4

ELECTRONICS PRODUCTION IN SINGAPORE

More foreign companies are upgrading their operations to bring product R&D, design, and other technology-intensive front-end activities into Singapore. Taking advantage of the growing sophistication of Singapore's skills base and industrial infrastructure, many multinational corporations (MNCs) are establishing CAD/CAM facilities and design labs to undertake electronic design and research.

The output of Singapore's electronics industry has been skyrocketing for the past three decades. In 1994, electronics output reached $35 (S$49.4) billion, which accounted for almost 50% of total manufacturing output. The output rose 24% from the previous year's $28.2 billion. [EDB 1995b] Output in 1995 was $41.6 billion. [Beh 1996]

Three key commercial groups that have been great assets in Singapore's electronics industry are multinational companies, high-growth local enterprises, and support companies. Singapore's strategic role to enhance their global competitiveness by taking advantage of MNCs. MNCs lend prestige to the country, enhance the economy by providing jobs, spin-off companies, and significant intellectual technology transfer. Some of Singapore's high-growth local companies have become the world's leading suppliers of their respective products. Creative Technology and Aztech Technology together produce almost 90% of the world's sound cards. [EDB 1995b]

The influx of investment by MNCs has led to development of vibrant support industries that offer a wide range of precision engineered parts, components, and services to support the manufacturing of disk drives, computers, color televisions, VCRs, instrumentation and other precision engineered products. In 1994, these supporting companies provided more than $800 million of goods and services to the industry. [EDB 1995b]

4.1 SEMICONDUCTORS AND PACKAGING

Singapore is a leading contract manufacturing center for surface mount technologies. Printed circuit board (PCB) assembly houses in Singapore now offer services from design to development, component sourcing to final product testing, and engineering support. The support industries have upgraded their technology to meet the challenges of miniaturization. New areas such as metal injection molding, PCMCIA connectors, and high-density multilayer PCBs have successfully been pioneered in Singapore.

4.1.1 Semiconductors

In Singapore only a handful of companies are involved in wafer fabrication, although more fabs are scheduled for build, especially to handle memory and ASICs. Today, Singapore's semiconductor sector is primarily involved in integrated circuit design, and automated assembly and testing. Some of the industry products and services that support the semiconductor sector are lead frames and wire bonding, burn-in and testing, wafer fab equipment, ceramic and epoxy packages, gas and chemical supplies, and cleanroom design and installation. [EDB 1995a]

Production of semiconductors, ICs, and other microcircuits was valued at almost $5.8 billion in 1994, about 17% of Singapore's total electronics production; output jumped over 60% from 1993 to 1994. Semiconductor devices held a similar share of total electronic production in 1995: output was $8.7 billion in that year. Since 1992, employment in this sector has been increasing at a compounded 7% rate, and the value added has increased at a compounded rate of 45%. [EDB 1995 a and b, and Beh 1996]

Some major companies involved in the semiconductor sector are given below, along with a description of their primary activities. [EDB 1995 a and b; Elsevier 1995 and 1996; ICI 1995 a and b; ICI 1996a]

- *SGS-Thomson Microelectronics*, the Franco-Italian chip manufacturer, is the only fully integrated semiconductor producer in the country. Its activities range from IC design and wafer fabrication to IC assembly and testing. The company announced the construction of a new plant in June 1996 with an investment of $400 million over the next four years. This plant will immediately double the company's existing $650 million investment in the region. SGS-Thomson focuses its R&D efforts on digital signal processing, IC design, advanced packaging, and wafer fabrication process development.

- *National Semiconductor (NS)* plans to invest around $130 million in its Toa Payoh plant, which would triple the facility's output. The Singapore branch produces and tests high-end ICs for applications in computing and communications. NS has announced the opening of a $212 million extension to the Singapore plant, which assembles and packages chips. The plant upgrade will be spread over five years.

SEMICONDUCTORS AND PACKAGING

- U.S. company *Harris Semiconductor* has opened a Center of Excellence in Singapore to combine functions critical to the Asian marketplace, such as design, technical support, market development, and regional sales.
- *Chartered Semiconductor Manufacturing* made its first functional 8-inch wafer using a 0.6 micron process at its new Fab 2 in Singapore. The company boasts of having the ability to reduce its circuitry linewidths to 0.35 microns and is currently working on reducing the geometries to 0.20 microns. CSM was established in 1987 and has worked as a silicon foundry for ASICs. It started construction of a Fab 3 facility in October 1995. [Beh 1996]
- *Siemens Components* of Singapore has established an IC design center involved with the production and application of chipcards for the Asian market. The company is building a fab in the region and has announced a $200 million investment to expand output in its Singapore assembly plant. The company performs R&D in Singapore in radio frequency, digital signal processing, and multimedia technology.
- *Toshiba* has entered into an agreement to license advanced semiconductor-related technology to one of Singapore's leading manufacturers, *Chartered Semiconductor*. The license covers transfer of process technologies for Toshiba's advanced 0.5 micron CMOS logic devices and will remain in force for five years.
- *Zoran Corporation* and *Goldtron Corporation* announced their joint venture, *Oren Semiconductor*, for the design and manufacture of "ghost" canceler ICs for the world's consumer television market.
- *Sharp* set up a new semiconductor design and development center in early 1995 at *Sharp/Roxy*, the firm's sales subsidiary in Singapore.
- U.S. chip maker *Advanced Micro Devices* plans to invest heavily in its Singapore facilities over the next three years.
- The *ESEC Group* of Switzerland is investing over $25 million in the next three years to set up an assembly facility in Singapore. ESEC makes automatic assembly equipment for the semiconductor industry.
- *Advance Systems Automation (ASA)*, a Singapore-grown company, is in the process of designing and developing an in-line manufacturing system for the semiconductor industry. The system will be housed in cleanroom containers to achieve mobility and to facilitate stacking into a multilevel production facility.
- *Hitachi* will work jointly with *Nippon Steel* and the *Economic Development Board (EDB)* to set up wafer processing plants for production of memory chips. The first plant's construction startup was announced in June 1996 with an initial investment close to of $1 billion. When completed, it will manufacture 64 Mbit DRAMs [Beh 1996].

- *TECH*, a consortium of *Texas Instruments, Canon, Hewlett-Packard, and the Economic Development Board*, will construct a 64-Mbit DRAM fabrication facility, using a 0.35 micron process. The new fab, costing $450 million, is the company's second and will raise TECH's monthly wafer output to 25,000. According to TECH management, the fab is expected to generate $600 million in revenues by 1999.

4.1.2 Passive Components and Display Devices

The primary products that come from Singapore's passive components and display devices sector are cathode ray tubes (CRTs), liquid crystal displays (LCDs), printed circuit boards (PCBs), capacitors, resistors, inductors, and quartz components. The companies involved in this sector mainly focus on product design and development, prototyping, process engineering, and high-volume manufacturing. Some of the industry products and services that support the sector are paper phenolic and glass-epoxy PCB laminates, precision metal parts, CRT glass panels, and process control system design and installation. [EDB 1995a]

The passive components and display devices sector experienced tremendous growth from 1992-1994 before it reached a plateau in 1995. This sector's production in 1995 was about $2.33 billion, compared to 1994 production of $2.1 billion. Employment in the sector has been stable over the same period, while value added has increased at a compounded rate of 24% — a good indicator of the increase in level of innovation in this sector. [EDB 1995 a and b]

Some major companies that are involved in this sector are given below, along with a description of their primary activities. [EDB 1995 a and b; Elsevier 1995 and 1996]

- One of the major production plants in the *Coilcraft Group* is located in Singapore. The company designs and manufactures surface mount inductors and transformers for the telecommunications industry.
- *Degussa Electronics* manufactures double-sided and multilayer PCBs used in computer systems and peripherals. The majority of the company's $57.1 million sales stems from its multilayer PCBs. The company plans to allocate $14.2 million to expansion projects over the next few years. It may begin to manufacture flexible PCBs.
- *Gul Technology* manufactures high-end double-sided plated through-hole and multilayer PCBs. It has been working jointly with *Teledyne* to manufacture rigid-flexible PCBs. The company will be spending $10.7 million in the next two years to increase production capacity and develop advanced interconnect substrates.
- Mixed signal ASIC designer *Silicon Systems* developed an advanced ultra-thin quad flat pack (UTQFP) that is an industry first: it is fully molded, which protects it from moisture and other environmental

elements. The Singapore design center has developed three LAN controller products and is developing sound card ICs for computers.
- *Murata Electronics'* Singapore facility has become the company's largest ceramic chip capacitor factory outside Japan. The company plans to invest $185.7 million in Singapore to upgrade production capacity and construct a new automated plant.
- The Singapore site for *Sony Display Devices* is Sony's largest TV picture tube production plant in the world. It earned $240 million in 1994. The plant, which makes color picture tubes (CPTs) ranging from 14 to 25 inches, is Sony's most productive CPT plant in the world. The division's new factory produces 25-in. and 29-in. CPTs. Altogether, the division's investment in Singapore has exceeded $354.6 million.
- *Mentor Graphics*, one the world's leading suppliers of electronic design [EDAS] automation systems, launched the industry's first design libraries program to support the entire PCB design process, including schematic capture, logic simulation, board layout, and analysis. The company introduced its Part Development System (PDS) to help customers simplify and streamline their efforts in part creation, assembly, and testing. Mentor Graphics has helped Singapore train technical staff for its electronics industry by installing over 300 of its systems in Singapore's two universities and four polytechnic institutes.
- *Sumitomo Bakelite* has opened a second epoxy molding compound manufacturing plant in Singapore to increase its total monthly output from 600 tons to 1,400 tons. The expansion makes it the world's largest facility in the field.
- *Hitachi Cable* is investing money into expanding production of IC lead frames, which consist of lead-on-chip (LOC) and normal lead frames.
- *Photronics, Inc.*, a leading U.S. photomask supplier, is setting up a plant in Singapore as the first phase of its globalization program.

4.2 PRODUCTS AND SERVICES

Product technology in Singapore is continually improving, as many local and multinational companies have established design and development activities there. Local companies have quickly taken advantage of the high governmental support by capitalizing on product development to increase market share. These companies are well versed in the art of producing modems, medical electronic products, VHF radios, burn-in test equipment, microcomputers, and applications involving microprocessors. [Elsevier 1996]

Electronic data processing equipment leads in electronics production. The sector produced over $19 billion of goods in 1995, accounting for almost 50% of total electronics production. In addition, the personal computer and computer peripherals sectors have gained ground. Local computer manufacturers are getting the world's attention with their competitive costs and highly reliable IBM compatibles. Disk drive exports alone have accounted for 39% of total computer equipment exports [Elsevier 1996].

4.2.1 Consumer Electronics

The main products in the consumer electronics sector are color TVs, audio equipment, VCRs, and key modules (CD pickups and video head drums). Singapore serves as the regional headquarters for many multinational corporations. Consequently, these corporations are greatly involved in R&D, technical support, and training. This sector's companies participate in high-volume manufacturing, product/process design and development, marketing and distribution, and parts procurement. The sector is supported by such products and services as automation equipment, fine-pitch SMT board assembly, contract product assembly, and precision plastic and metal tools, dies, molds and part makers. [EDB 1995a]

Production of consumer electronics has shown great stability in terms of employment and value added over the last four years. Production has enjoyed a compounded average annual growth of 10%. Output was $3.9 billion in 1992, $4.1 billion in 1993, $5 billion in 1994, and $5.2 billion in 1995. With employment remaining the same, it appears that the sector's production efficiency has improved. [EDB 1995a, Elsevier 1996]

Some major companies that are involved in this sector are given below, along with a description of their primary activities. [EDB 1995 a and b; Elsevier 1995 and 1996]

- *Aiwa* spent $17.7 million to expand and automate its audio equipment manufacturing operations and to set up an R&D center in Singapore. The R&D center will house an electromagnetic compatibility and immunity laboratory. The lab will allow Aiwa to ascertain the level of electromagnetic interference susceptibility and immunity its products have. Aiwa Singapore currently manufactures audio-visual products and is the parent company's largest offshore manufacturing facility, accounting for over 40% of worldwide output.
- *Matsushita Electric Industries* boasts one of the most automated consumer electronics plants in Singapore. The plant designs and manufactures a range of audio electronic products, including cassette and CD mechanisms. It is also the home of the AV/Information Research Center, which performs research on audio-visual compression and processing algorithms.
- *Philips Singapore* designs and manufactures color television sets, audio products, tuners, and domestic appliances. The company recently financed $27.1 million for a new building at its Toa Payoh site for a color TV R&D center. Philips' R&D centers designed the Monster Micro mini-compo audio system, which has been a great success.

PRODUCTS AND SERVICES 51

- *Sony's* Singapore site produces parts and key modules such as video drums and magnetic head cylinders, and CD optical pickups and CD mechanisms. The company has undergone a $15.6 million expansion to make electron guns used in Trinitron picture tubes, its first fully integrated electron gun line outside Japan with the ability to assemble complete electron guns from raw materials.
- Singapore's Electronic Devices division of *Hitachi* has established an R&D center in the country that will enable it to become more self-reliant for engineering support. The R&D center will develop such products as the Wide Vision CPT.
- *Thomson Consumer Electronics* set up a $35.3 million joint venture with *Toshiba Singapore* to design and manufacture VCRs in Singapore. Toshiba Singapore is currently one of the largest TV set assemblers in the country.
- *P-Serv Technologies* has developed Serv-Touch, a compact home automation system for appliances. The system will be able to control lighting, VCRs, and air conditioners through radio frequency, thus requiring no extra cabling.
- *Taiyo Technology*, producer of parts for switching devices in audio-visual products through two-color injection molding and laser etching, has expanded its Singapore operations by building a new factory after doing business for less than two years.
- *Robertson Audio*, one of the few niche producers in Singapore, produces state-of-the-art audio amplifiers that sell from $5,000 to $6,000.
- U.S. home appliance manufacturer *Whirlpool* will invest $7.1 million in Singapore to set up an R&D center and an Asian operational headquarters. The R&D facility will focus on design and development, prototype production, and product testing for the company's air conditioners, refrigerators, and washing machines.
- *Varta Batteries* has released a prismatic nickel hydride battery for the cellular telephone market. The batteries' low profile meets the demand for cellular phone miniaturization.

4.2.2 Data Storage

The data storage sector's main products are rigid disk drives, removable cartridge drives, tape drives, CD-ROM drives, and recordable compact disc media. The key activities for this sector are high-volume assembly of precision devices, manufacturing process enhancement, device continuation engineering and cost redesign, and subassembly design. The sector's technical capabilities include test/ECC microcode development, servo electronics development and redesign, and precision electromechanism engineering. The sector is supported by products and services such as cleanroom design and installation, automation systems design and development, metal casting and machining, fine pitch SMT board assembly, and rapid prototyping. [EDB 1995a] Singaporean disk drive manufacturers assembled more than 29 million rigid disk drives in 1994, accounting for 46% of global production.[EDB 1995b]

The data storage sector is the clear leader in Singapore's electronics industry in terms of the monetary value of its output. From 1992 to 1994, production in the sector increased from $6.4 billion to $7.7 billion, a compounded annual average growth rate of 14%. Employment in the sector has shown the same steady growth in the 1990s, and the data storage sector has become the sector with the highest employment. One downside of the growth in this sector, is that the value added dropped from $1.7 billion in 1992 to $1.5 billion in 1993; nevertheless, the value added rebounded in 1994 to reach approximately $1.8 billion. [EDB 1995a and b; Elsevier 1995 and 1996]

Some major companies that are involved in this sector are given below, along with a description of their primary activities. [EDB 1995 a and b; Elsevier 1995 and 1996]

- *IBM Storage Systems Division* manufactures high-end rigid disk drives with at least one gigabyte of storage capacity. In addition to supplying IBM itself, the plant supplies to the open market.
- *Kotobuki Electronics* uses state-of-the-art automation equipment to manufacture rigid disk drives for *Quantum*.
- *Optics Storage* is developing an 8X CD-ROM drive, the fastest in the industry today. The company is looking into developing a quad-speed CD-recordable drive for audio and data.
- *Mitsubishi Chemical* is investing $17.9 million to manufacture CD-recordable media. It is the first high-volume manufacturing facility outside of the United States or Japan to produce this product.
- *Seagate*, a world leader in the disk drive industry, has built a $130 million facility in Singapore where it will assemble disk drives and make printed circuit boards using state-of-the-art chip on board technology. This is on top of existing investment of more than S$800 million. In addition, Seagate is investing $142.8 million in two new disk media plants. [Beh 1996]
- *Western Digital*, the fourth largest producer of rigid disk drives by unit volume in the world, uses its strong IC design expertise to retain a time-to-market lead over its competitors.
- Disk media company *StorMedia* opened a $40 million three-line plant in June 1995 and is reported to be considering opening a second facility.
- Japan's *Hoya Corporation* has invested $30 million in the first phase of its project to produce rigid disk media on glass substrates.
- U.S. firm *Conner Peripherals* has announced multimillion dollar investment plans in the region. Initial agreements have already been signed.

PRODUCTS AND SERVICES   53

- *Varta Batteries* developed a rechargeable button cell nickel hydride battery for the memory backup market. The breakthrough product won the German Innovation Award silver medal in 1994. The battery is environmentally friendly since cadmium is not used. Varta spends about $2 million a year on R&D.
- *Philips Singapore* has set up *Laser Optics Asia (LOA)* to act as a competence center for optical storage devices. The center will focus on advancements in optical storage for CD players, CD-ROM drives, and optical disk drives. LOA has established an automated production plant for CD/CD-ROM key modules and components.

4.2.3 Computer Systems

The computer systems sector develops a variety of products, including desktop personal computers, notebook PCs, mobile computers, high-end servers, multimedia products, and motherboards and related subassemblies. Primary activities that this sector pursues are high-volume manufacturing, design and development, process engineering, and marketing and distribution. Technical capabilities in Singapore include human interface technologies, digitization techniques, miniaturization, and new product management. A typical company in this sector would be supported by contract manufacturing, automation equipment, precision plastic/metal parts and electronic components, and rapid prototyping services. [EDB 1995a]

The computer systems sector is merely a shade behind the data storage sector in terms of monetary value of output. This is due to the sector's dramatic increase in output from 1992-1994. The sector's output was $5.0 billion in 1992 and grew to just under $7.7 billion in 1994. Employment has followed the same growth trend, employing 5,500 people in 1992 and 8,300 in 1994. [EDB 1995a, Elsevier 1996]

Some major companies that are involved in this sector are given below, along with a description of their primary activities. [EDB 1995 a and b; Elsevier 1995 and 1996]

- Singapore's *National Science and Technology Board (NSTB)* has announced another 4-year R&D partnership with *Apple Computer*, which will invest over $28 million in the region. Apple has been operating in Singapore since 1981, and Singapore has Apple's global charger to design, develop, and manufacture entry-level MACs. [Beh 1996] The original plant produced Apple II products for the Pacific region and expanded to manufacture the Macintosh Classic and Macintosh LC. Apple has combined resources with Singapore's *Institute of Systems Science* to develop human interface technologies such as speech and handwriting recognition in multiple languages. The plant operates 24 hours a day with a staff of more than 500.
- *Compaq's* Singapore facility manufactures desktop and notebook PCs, power supplies, PCBAs, and related subassemblies. The plant employs

- *IPC Corporation* produces desktop and notebook personal computers, point-of-sale systems, and multimedia products. The company has recently invested $17.9 million in a second plant to expand manufacturing output. IPC has become one of the largest sellers in the Singaporean PC market, second only to Compaq. This company's five-year R&D program is geared towards human interface, signal processing, wireless communications, and multiprocessor technologies.
- The Singapore-based MNC *Creative Technology* is the world's leading supplier of sound cards. The company designs and manufactures sound and multimedia products for IBM compatibles as part of the Blaster series. Creative's SoundBlaster platform allows PCs to produce high-quality audio applications and has been accepted as the industry standard in sound platforms for PC-based software. The company's R&D team works on developing sound and video cards, PC video conferencing, and other multimedia products.
- *Aztech Systems of Singapore* is the world's second largest producer of add-on multimedia cards. The company has just designed and developed a PC-based audio telephony board that consolidates fax, modem, voice, and telephony functions. The company has developed expertise in digital and analog design, mixed-signal ASIC design, microcontroller programming, and device driver development. Aztech plans to focus future R&D on DSP techniques, video conferencing technology, data compression, wave table synthesis sound generation, and CD-ROM technology.
- Singaporean engineers at *Hewlett-Packard (HP)* have designed a powerful graphing calculator (HP38G) for worldwide markets. Two other products to be launched by HP Singapore are an innovative high-end organizer and a personal communicator. HP's Singapore product line includes calculators, palmtop computers, inkjet printers, ICs, LED components, personal computers, and network products. HP has recently received the Distinguished Partner in Progress award from the Singapore government for 25 years of commitment to the country. The company is building a new $73 million facility to accommodate PC manufacturing and assembly and testing of high-end ICs.

4.2.4 Telecommunications

Singapore's telecommunications sector produces pagers, cordless and cellular phones, communications antennas, navigation products, and two-way radios. The sector's key activities are product design and development, R&D, manufacturing, and marketing and distribution. Its technical capabilities include digitization, fine-pitch SMT assembly, miniaturization, automation, and software development. The industries that support the telecommunications sector are ASIC design, hybrid circuit fabrication and assembly, semiconductor wafer assembly and testing, electromagnetic testing, and precision plastic and metal working. [EDB 1995a]

The country's telecommunications sector has shown signs of inconsistency. From 1992-1993, output jumped from $0.9 to almost $1.4 billion, an increase of almost 50%. From 1993-1994, however, the sector's output dropped to $1.2 billion. Employment and value added in the sector followed the same trend over the same time period. Nevertheless, the telecommunications sector has enjoyed a compound average annual growth of 14%. [EDB 1995a; Elsevier 1996]

Some major companies that are involved in this sector are given below, along with a description of their primary activities. [EDB 1995 a and b; Elsevier 1995 and 1996]

- *Motorola* is investing $30 million to consolidate its paging operations in Singapore. Motorola plans to dive into the field of digital signaling with intelligent networks having services such as call forwarding. Singapore is Motorola's Asia Pacific headquarters for Paging Subscriber, Cellular Subscriber, Semiconductor Products, and Telepoint System divisions; it is also home to the Land Mobile Products division. Motorola has invested over $200 million in Singapore and plans to spend another $100 million in the next five years.
- *Trio-Kenwood*, a wholly owned subsidiary of U.S. corporation *Kenwood*, manufactures wireless communications products that account for 30% of its global communications devices. The company has established an R&D center for audio and communications products. Trio-Kenwood also services the consumer electronics sector by manufacturing hi-fi and automobile audio systems. It is Kenwood's largest plant.
- Telecommunications products manufacturer *Uraco Technologies* has developed a solution to the telephone line shortage problem that does not involve laying new cables. Uraco designed a digital multiplexer, the DigiPlex U6, that can increase an existing phone system's capacity by six times. The system is installation-ready for most subscriber telecommunications systems.
- *Philips* announced in June 1996 that it will build in Singapore its only Asian facility to manufacture telephones.
- Singapore-based *Powermatic Data Systems* produces a wide range of LAN products including adapters such as Arcnet, Ethernet, and Token ring, and interconnectivity products such as hubs and concentrators, network workstations, file servers, and mass storage subsystems.
- *Rockwell Defense Electronics*, through its *Collins Avionics & Communications Division*, has set up a Pacific Rim service center in

Singapore. The center provides product repair and technical support for defense communications and navigation products.

4.2.5 Office Automation

Singapore's office automation equipment sector develops inkjet, thermal, and impact printers; facsimile machines; personal word processors; and printer consumables such as head cartridges. The sector's key activities are high-volume manufacturing, product design and development, and process engineering. Its technical capabilities in Singapore are precision electromechanical engineering, data and image compression, and specialized font development, aimed at Asian markets. The sector is supported by such products and services as plastic and metal working, fine-pitch SMT board assembly, automation equipment, and contract manufacturing. [EDB 1995a]

Office automation has been a booming sector of Singapore's electronics industry. From 1992-1994, output increased from $1.7 billion to $3.6 billion, a compounded increase of 47%. Employment has increased at a compounded rate of only 6%, which indicates either that production efficiency has increased or the level of innovation has gone up, or both. The value added enjoyed a growth trend similar to that of output. [EDB 1995a; Elsevier 1996]

Some major companies that are involved in this sector are given below, along with a description of their primary activities. [EDB 1995 a and b; Elsevier 1995 and 1996]

- The Asia Peripherals Division of *Hewlett-Packard* exercises complete control over the company's portable printer business. Since the division's establishment, it has developed the first Deskjet printers for the Asian market and the Deskjet Portable, which weighs just 4 pounds. This division is one of the first enterprises to start design and development of inkjet printers in Singapore.
- *Matsushita Graphic Communication Systems* in Singapore is the largest maker of fax machines in the world, controlling over 30% of the global market. The company produces a wide variety of fax machines using thermal, inkjet, and laser printing technologies. Some of the company's other products combine fax, printing, and copier functions.
- Even as a late entrant to the inkjet printer field, *Seiko Epson* has established a name for itself. Its major strength in the field is in its proprietary inkjet printhead, which allows for precision-controlled printing.
- *Printronix Inc.*, a U.S. manufacturer of printers for mid-range computers, will manufacture the company's latest family of high-speed line-matrix printers, the P4XXX.

PRODUCTS AND SERVICES 57

4.2.6 Contract Manufacturing

The contract manufacturing sector provides a great deal of support to most of the other sectors in Singapore's electronics industry, especially the data storage, personal computer, and peripherals sectors. The sector's main services are PCB assembly for various electronic products and comprehensive electronic systems such as printers and computers. Contract manufacturers for electronic companies perform high-volume turnkey manufacturing, PCB layout, product testing, process engineering, and product redesign for manufacturability and cost reduction. The sector's technical capabilities are surface mount, chip on board, and chip on flex technologies. [EDB 1995a]

Output in the contract manufacturing sector did not grow from 1992 to 1993, but did rise 13% from 1993 to 1994. On the other hand, employment in the sector has increased steadily. The sector thrives on servicing small corporations that do not have the physical size or manpower to operate their own manufacturing facilities. One problem with the sector is the lack of growth in value added since 1992, which has held steady at approximately $170 million. [EDB 1995a]

Some major companies that are involved in this sector are given below, along with a description of their primary activities. [EDB 1995 a and b; Elsevier 1995 and 1996]

- *SCI Systems'* Asia Pacific hub for manufacturing as well as RHG is in Singapore.
- Singapore is *Flextronics'* core location for Asian operations. The company provides design and manufacturing services for technologies such as chip on board, multichip modules, and fine-pitch surface mount technology. The company is working with GINTIC to increase proficiency in flip chip on glass technology.
- The *Goldtron Electronics* branch of the Goldtron Group maintains a high degree of diversity in comparison to many electronics businesses. Some of its new products are LetsTalkFax, the world's first "talking" fax; cellular phones using the ETACS and EAMPS systems; Proteq, a failsafe security and automation system; Boardwizard, a diagnostic testing device; and touch screen multimedia systems.
- *Natsteel Electronics*, the second largest printed circuit board assembly company in Singapore, is now working with GINTIC to master BGA and flip chip on glass technologies. The company sees these technologies being in demand because of the drive towards miniaturization.
- *Venture Manufacturing* has gained a great reputation for building quality products at low cost. Its products cover many segments of the electronics industry, including telecommunications equipment, security systems, digitizers, navigation systems, printers, and multimedia

products. Two of its proprietary products are *Jetwriter*, a portable inkjet printer; and *Brushwriter*, a Chinese word processor with an English-Chinese dictionary.

- *Delco Electronics Singapore* has set up an international procurement office (IPO) and an advanced manufacturing engineering division in Singapore to develop and integrate the next generation of manufacturing processes and packaging to support its future needs. The company manufactures engine control modules, audio systems, pressure sensors, and voltage regulators for worldwide markets. It has 2,500 employees.

4.2.7 Applied Research and Development Centers of Excellence

The swift growth of a well-trained technical workforce in Singapore has allowed companies to expand their operations to include R&D in addition to manufacturing. In 1993, the electronics industry's R&D sector devoted a record $268 million to its activities, or 38% of the country's total R&D spending. Singapore's readiness to grasp the future of electronics technology is evidenced by the increasing number of research scientists and engineers (RSEs). In 1993, RSEs numbered well over 2,000, accounting for one-third of the total number of RSEs in manufacturing. In addition, the annual expenditure per RSE equals $124,000 exceeding both Taiwan's and Korea's values [EDB 1995b].

Aside from corporations integrating R&D into their operations, Singapore has a strong base of research centers designated "Centers of Excellence." Each one focuses on a specific area of electronics and serves to train manpower, improve core technologies, and facilitate technology transfer in their respective areas of focus. The Centers of Excellence that work with the electronics industry are described below. [EDB 1995a; Elsevier 1996]

- *The Magnetic Technology Centre (MTC)* was started in 1992 to aid data storage and magnetic products businesses in Singapore. Now called the *Data Storage Institute*, it currently focuses its R&D on mechanical and electrical systems design; heads, media and materials; and optical systems. The research center has approximately 60 researchers.
- *The Institute of Microelectronics (IME)*, founded in 1991, involves itself in precompetitive applied R&D in the following areas: microelectronics systems applications, VLSI design and testing, microelectronic process technology, failure analysis and reliability, and advanced packaging development. The IME looks to fortify the semiconductor industry in Singapore.
- The *Centre for Wireless Communications (CWC)*, which works with the telecommunications sector, was established to promote development of all facets of wireless communications. The CWC works jointly on R&D projects with companies, trains manpower, develops core technologies, and fosters technology transfer. It is currently working with Motorola

on developing two-way calling CT2 systems, and with Singapore Telecoms on developing digital cellular mobile radios.
- The *GINTIC Institute of Manufacturing* is the foremost institution in Singapore for the development of intelligent and precision manufacturing. Currently, its five main research programs are advanced precision manufacturing, near net shape manufacturing, printed circuit board assembly technology, surface technology, and automation technology.

4.3 CASE STUDIES

In order to gain more of an appreciation for Singapore's prowess in the electronics industry, it is useful to dissect some of the country's successful corporations and government organizations.

4.3.1 NatSteel Electronics

Singapore's NatSteel Electronics was established in 1981. From 1992 to 1995, the company's revenue grew by more than 80% annually from $80 million to $500 million. Its 80,000 ft^2 plant employs 750 people and has 12 SMT lines. The facility was ISO 9002 certified in 1993. From Singapore, Natsteel Electronics provides infrastructure support for all other NatSteel facilities in East Asia.

NatSteel performs contract manufacturing for both low- and high-volume products. This can be done because its Singapore facility has two separate floors to manage the different volume levels. The low-volume lines have a screener, an MV2C chip placer, an MPA3 module placer, IR, inspection, ICT, and functional testing capabilities. Complete line product changeover can be done in two hours, and sample parts can be made in the same amount of time before production. Statistical process control is used for each operation.

The facility's high-volume lines have the capability for screens, manual plated through hole (PTH) placement in paste, connector insertions, MV2C chip placement, MPA module placement, IR, cleaning, integrated circuit test (ICT), burn-in, functional testing, and visual inspection. The company reaches a 96% yield in ICT and 98% yield in functional testing.

The process engineering department of NatSteel works jointly with GINTIC in Singapore and Panasert in Japan. Its work with GINTIC revolves around process technology development. Its engineers are qualifying tape carrier package (TCP) attachments with a Panasert machine and qualified flip chip on glass using stud bumping bonding and anisotropic conductive films. They have qualified and built products with organic coating, 15 mil QFPs, and TABs.

NatSteel's customers involved with computers and peripheral cards include BMI, Chicony, Compaq, Diamond Multimedia, Hewlett-Packard, IBM, JTS, Multiwave, Video Logic, and Western Digital. Its customers in the telecommunications field include Nortel, Hewlett-Packard, Supra, and Yupiteru. The company does consumer electronics work for Delta, Giken, Hipro, Hewlett-Packard, Liteon, Sumitomo, and TEAC. NatSteel is also involved in manufacturing RF electronics for General Instruments and Wireless Access. [Kelly 1996]

4.3.2 SCI Manufacturing

SCI originated as an R&D company in 1961 and entered the contract manufacturing business in 1977. Since then it has become the largest contract manufacturer in the world. SCI management has a firm belief in establishing company facilities close to customers and has consequently started 22 facilities globally. The firm's worldwide revenues have been over $2 billion the last two years and are on track to reach $4 billion in 1996. The Singapore facility alone generated $330 million in revenue in 1995.

SCI Singapore has 18 SMT lines that use Fuji equipment. Some of its testing equipment includes products from Hewlett-Packard, Genrad, Zentel, and Fairchild. The company has a 85-90% ICT yield and 92-95% functional testing yield. The Singapore facility also does failure analysis at the component level. It makes use of GINTIC's local facilities and expertise in the area and obtains local perspectives and networking from local manufacturers.

Some of SCI's customers are Hewlett-Packard, Seagate, Singapore Technologies, Compaq, Texas Instruments, Exabyte, NEC Semiconductors, Baxter, and Apple. Its wide range of assembled products includes PC motherboards, SIMMS, LANs, printers, disk drives, modems, tape drives, and video tuner cards. [Kelly 1996]

4.3.3 Venture Manufacturing

Venture Manufacturing, established in 1984, is considered one of the fastest growing electronics companies in the region. The company became public in 1992 and experienced an 83% growth from 1993 to 1994. Venture Manufacturing has two ISO 9002 certified facilities in Singapore, one in Ang Mo Kio and another in Kallang. The company's goal is to provide cost-effective manufacturing for its customers while providing other value services such as concurrent engineering and product prototyping.

Venture uses Fuji placement equipment and currently has over 20 SMT lines. Some of its lines contain a DEK265 screener, Fuji CP-4 chip placement machines, Fuji IP-II module placement machine, and Heller 1500 or BTU reflow ovens. The lines are hard-coupled and use NUTEK board-handling equipment, which is manufactured in Singapore.

One component of Venture's strategy is customer and product diversification. Some of its principal customers are Hewlett-Packard, Apple, Sony, Compaq, Iomega, and Adaptec. It has manufactured a wide range of products, including bar code printers, portable printers, and GPS systems.

Each new employee at Venture goes through a 2-day general training session, followed by on-the-job training for specialized operations. Currently, 30% of Venture's labor force comes from Malaysia, and the company has a very low attrition rate. All of the employees are involved in a quarterly profit-sharing plan. [Kelly 1996]

4.3.4 Chartered Electronics Industries

Chartered Electronics, a member of Singapore Technologies, was established in 1980 with government-sponsored venture capital to design and develop video cards for personal computers. Since then the company has expanded its business to include all PCBs that have a high labor content. It has a global customer base and manufactures products for the telecommunications, industrial, PC, and office equipment sectors.

Chartered's strategy is to shift to complete product production and to provide higher-value-added services. It has already exhibited the ability to provide complete product production, having produced a portable language translator. The company had a product redesign activity where its engineers performed the redesign themselves, providing their customer with a 20-30% cost reduction. Chartered works closely with GINTIC and other government resources to decrease its process development cost.

Chartered uses KME placement equipment on its SMT lines and has a cycle time of 3-4 days. The labor force includes Malays and Chinese in addition to Singaporeans. The turnover rate is approximately 10% per year. [Kelly 1996]

4.3.5 ST Assembly Test Services

ST Assembly Test Services (STATS), one-quarter of the semiconductor group of Singapore Technologies, is a one-stop service shop for the assembly and testing of ICs. STATS was established in January of 1995, and assembly began in August of that year. Its first profitable month came only twelve months from inception, in December 1995.

STATS currently leases over 90,000 ft^2 of space and will occupy another 500,000 ft^2 facility within the next year. It is considering building a prototype facility in the United States. The major shareholders of STATS include the EDB, Singapore Technologies, and Japan's Seiko-Epson. The capital plan is $317.75 million, and STATS currently has $76.25 million of equipment installed.

Seiko-Epson takes on the responsibility of technology transfer, obtaining packaging capacity, and preferential pricing. The key managers of STATS have many years of accumulated experience in companies such as TI, Fairchild, National, and AMD. The company employs almost 600 people and has a customer base of over 20 electronics companies.

STATS has the capability of offering its customers reduced cost and shorter cycle times. It has provided a 2.5-day turnaround on product samples at no extra cost to the customer. Much of the company's new capital investment is geared towards obtaining highly automated and faster equipment that will provide better quality and shorter cycle times.

Currently, STATS' packaging focus is on higher value added, and higher pin count packages of QFPs and PLCCs. Future plans in packaging are to include assembly and testing of BGA, TQFP, and TSOP devices. [Kelly 1996]

4.3.6 Texas Instruments (TI)

Texas Instruments' Bendemeer IC assembly and test facility was established in 1968 to package transistors. The company's strong desire to set up a plant in Singapore is shown by the fact that it started production only 50 days after signing an agreement with the government of Singapore. The plant has grown dramatically since then, and it now produces most of TI's DRAM products. Initially, packaging technology was imported from the United States and the actual packaging was done in Singapore. Now all TI DRAM packaging responsibilities are handled in Singapore.

The Singaporean IC facility is extremely capital intensive, having $1.1 billion in total assets and an annual investment of $55 million. The company has a competitive edge in IC packaging cost, even with expensive Singaporean labor. This is because of the higher worker skills, which permit use of highly automated equipment and engender higher yields, shorter times to yield, and reduced process and testing times.

TI Singapore's devotion to customer satisfaction is exhibited strongly in its 99.5% on-time delivery record for the first quarter of 1996. The company focuses on three factors in the improvement of IC packaging technology: miniaturization, density improvement driven by component integration, and development of IC packaging at the assembly site. The company uses the Institute of Microelectronics' failure analysis facilities for quick turnaround on sample production. TI has not had a packaging-related qualification failure since 1989. [Kelly 1996]

Chapter 5

ELECTRONICS PRODUCTION IN MALAYSIA

In order for Malaysia to become an industrialized nation by the year 2020 as it plans to do, its R&D facilities must progress in tandem with the massive investments made in the manufacturing sector. The Malaysian government and private sector are aware of this and have already made inroads into the world of R&D. Below are examples of research projects and other activities undertaken by major electronics corporations in Malaysia.

5.1 SEMICONDUCTORS AND PACKAGING

Malaysia has been an important location for electronic packaging and equipment manufacturing for several decades. As of late, however, the Malaysian government has identified semiconductor manufacturing as one of its focus areas. Through the Vision 2020 plan, the government has stated its desire to increase R&D in semiconductors by offering incentives to existing businesses. Currently, semiconductor output totals approximately $8 billion[1] each year, over 40% of Malaysia's electronics sector. [Dunn 1995]

Malaysia is the world's number three producer of and leading exporter of finished semiconductor chip devices. Since the early 1970s, the semiconductor devices sector has grown by approximately 20% annually. The growth is evidenced by the presence in Malaysia of global electronics giants, including Intel, Texas Instruments, Advanced Micro Devices, and Hewlett-Packard.

Discrete semiconductors and integrated circuit packaging hold the number one position for Malaysian electronics production, accounting for over 25% of total electronics production in 1995. Discrete semiconductors and ICs are Malaysia's leading electronics exports. Some examples of the enormous investments in semiconductor manufacturing and packaging by

---

[1] All $ in this chapter are $US unless otherwise specified

firms in Malaysia are given below. [Dunn 1995; Elsevier 1995 and 1996; ICI 1996, a and b]

- *Intel Corporation*, one of Malaysia's largest hi-tech investors, has production and R&D facilities in its Penang factory. The R&D Center is now designing 8-bit and 16-bit very large scale integrated (VLSI) microcontrollers for broad market applications. It currently has 200 engineers in the factory and spends about $15 million annually on R&D.
- *InterConnect Technology*, recently founded by the Malaysian government, is the country's first state-of-the-art semiconductor manufacturing facility. InterConnect's mission is to produce semiconductor products for customers worldwide. The facility, which produces 25,000 8-in. wafers per month, is equipped with 0.35 μm equipment, and has a roadmap to achieve 0.25 μm.
- *Motorola* has increased its monthly capacity in its Seremban wafer fab to 4,000. The fab uses a 3 μm bipolar process on 4-in. wafers. Sixty percent of the wafers are used by Motorola's own assembly and test facilities in Malaysia.
- *Kedah Wafer Emas*, a Malaysian company owned by Taiwan-based *Hualon*, is at work on a new wafer fab costing $800 million. The fab will use 0.5 μm CMOS technology to produce 24,000 wafers per month.
- Malaysian IC packaging firm *Unisem* experienced a 60% growth from 1993 to 1994. The majority of Unisem's business comes from the United States. The company doubled its capacity from 1994-1995 and is expanding into the area of final testing.
- *The Malaysian Institute for Microelectronic Systems (MIMOS)* will be building Malaysia's own microchip plant at a cost of $44 million. The plant will be functional in 1996 and will produce chips for domestic needs. MIMOS will set up a Microelectronics Center in the Kulim Hi-Tech Industrial Park. It will be ready in 1997.

5.2 PRODUCTS AND SERVICES

While Malaysia has placed heavy emphasis on semiconductor manufacturing, the consumer electronics and electronic data processing (EDP) sectors have begun to steal some of the spotlight. One facet of the Vision 2020 plan is to increase Malaysia's diversification in the electronics industry so that the country can act as a comprehensive production and supply base for global markets. The electronics industry has displayed increased depth with the influx of foreign small- to medium-sized companies.

This has resulted in an increased drive for those in the semiconductor industry who have moved from simple assembly and testing to setting up wafer facilities. Another beneficial consequence is the cohesiveness built among electronics firms in different sectors that must interact with each

PRODUCTS AND SERVICES 65

other during product development. The only disadvantage to this rapid growth is in itself an excellent indicator of how well the industry is doing: the government along with Malaysian firms are now worried about land availability and the long-term labor supply. [Elsevier 1996]

Malaysia is currently well situated to exploit GATT in order to increase diversification in its electronics industries. Semiconductors represent a significant cost for any electronic product. With strategic diversification in areas of flat panel displays, molding, and circuit boards, system supplies can avoid import duties.

Two of the fastest-growing sectors of electronics exports are the audio and visual (AV) equipment and electronic data processing (EDP) sectors. The AV equipment sector has had a compound average annual growth (CAAG) exceeding 15% over the last five years. The EDP sector experienced an impressive CAAG of 35% over the same period.[Elsevier 1996]

Aside from corporations, the electronics industry has seen aid from organizations established by the government. MIMOS and the Standards and Industrial Research Institute of Malaysia have placed an enormous emphasis on the research and development of electronics and electronics systems. Some recent investments in the consumer electronics sector that Malaysia has seen from both corporations and the government are described below. [Elsevier 1995; Elsevier 1996; ICI 1996, a and b]

- Engineers at *Hewlett-Packard Malaysia* work closely with R&D teams in San Jose in areas ranging from product development to equipment purchasing. Over the next few years, the Malaysian R&D center will contribute 25% of HP's worldwide R&D. In addition, HP has transferred its computer disk drive manufacturing facility from England to Penang, making it the only factory outside the United States to manufacture the product.
- *Komang USA*, the world's leading manufacturer of hard disks, has an R&D facility in Penang that employs 80 engineers. They are working on a thinner, lighter, and cheaper hard disk with faster access time. Komang has invested $100 million to produce a new 500-megabyte 8.7 cm disk, making Malaysia the third country in the world after the United States and Japan to manufacture the product. Komang plans to make the Penang plant into the regional manufacturing center and headquarters for its high-end products.
- *Motorola* has developed a walkie-talkie for lowband VHF and UHF in its Penang R&D facility. The Penang center contributed to the development of the CT2 product enhancement program for subscriber handsets and public base stations. Motorola's R&D activities in Malaysia include product design and development.
- *Eng Technology Holding Berhad*, a Penang-based Malaysian company, has been contracted to manufacture E-block rotor assemblies, a

component of hard disk drives, for *Quantum Corporation*. Its subsidiary *Eng Technologi Sdn. Bhd.* was contracted by *Micropolis Corporation*, a major U.S. high-capacity disk drive producer, to perform R&D on two of its products.

- The German firm *Robert Bosch* has invested heavily in R&D in its Penang factory. The facility is to design new car stereo components and electronics for controlling windshield wipers. It currently has 32 engineers in Penang.
- *Sony* has opened an R&D center in Selangor, its third outside Japan. The other overseas R&D centers are in Britain and the United States. The Malaysian center was established to design television sets and other audio-visual products.
- *Hitachi Electronic Products* designs and develops video cassette recorders in its Bangi R&D center.
- *Sapura Holding Group* produced the world's first voice-activated telephone in its R&D center.
- *Sharp Corporation* of Japan has invested in its largest television manufacturing plant outside Japan through *Sharp/Roxy Electronic Corporation Sdn. Bhd.* It is also investing $40 million to build an R&D center in Jahor to design and develop new TV models.
- *Grundig Sdn. Bhd.* spends $1 million per year on R&D to develop new audio-visual and electronics products. Its R&D activities include circuitry drawings, engineering and design, prototype production, and testing and evaluation. The R&D center was set up at a cost of $16 million.
- *Computer Resources Manufacturing* has invested $80 million in R&D to manufacture pocket hard disks. These are portable disks that store up to 1 gigabyte of data compatible with IBM file servers, personal computers, and notebooks. They work with MS-DOS, Microsoft Windows, and Novell Netware.
- *The Standards and Industrial Research Institute of Malaysia (SIRIM)* will be setting up the Advanced Materials Research Center in the Kulim Hi-Tech Industrial Park. The center will perform R&D on advanced materials such as ceramics, composites, and new materials for aerospace and semiconductor applications.
- *IBM World Trade Corporation* and *MIMOS* are collaborating on R&D programs in microelectronics and information technology. MIMOS recently received a Power PC Development Kit to explore the development of applications relevant to Power PC technology.

5.3 CASE STUDIES

In order to gain more of an appreciation for Malaysia's prowess in the electronics industry, it is useful to dissect some of the country's successful corporations and government organizations.

5.3.1 Intel-Penang

Intel's Penang facility, established in 1972, was its first plant outside the United States. Its primary focus at the time was producing low-cost, labor intensive, low-technical-content products. Over the last two decades, Intel-Penang has grown into a highly automated, capital intensive IC assembly and test facility. Its growth is the result of many initiatives, one of them being the addition of testing activities in the late 1970s. In the 1980s, Intel added a customer warehouse and process and product capabilities. Intel-Penang has now shouldered the customer interface and R&D responsibilities for microprocessor product packaging, processing, and silicon design capability supported by CAM.

The 60-acre facility in Penang houses 3,700 employees and has $800 million of capital investment, up from $4 million when it started in 1972. The Penang facility has about a 1% monthly attrition rate, down from over 30%. Decreased turnover is due mainly to newly founded corporate programs such as profit-sharing, employee stock options, work recreation programs, and Malaysia's first company-sponsored kindergarten.

The Penang facility manufactures Intel's 80486DX, 80486DX2, Pentium, and Pentium Pro microprocessors; 8-bit, 16-bit, and 32-bit microcontrollers; Pentium core logic chipsets; and communication devices. It also supports various package types from large pin count ceramic PGAs to QFPs, PLCCs, and PDIPs.

Intel-Penang is the most productive facility within the whole Intel organization. This is due to its 20-year investment in research and development and the character of the employees. The workforce is known to have high work ethics and a desire to learn. With their lead in assembly technology, Intel will continue to deploy their latest packaging technology in Malaysia for the foreseeable future.

5.3.2 Motorola Microcontroller Technology Group - Kuala Lumpur

The Motorola facility in Kuala Lampur is the company's largest IC packaging facility, employing nearly 5,000 people. Motorola has been performing manufacturing operations at the Kuala Lumpur site for 22 years. Its other Malaysian sites are in Penang, Serban, and Philips JV. The Kuala Lumpur facility is Motorola's flagship for reliability and quality while sustaining competitive yields for complex package types.

The highly automated Kuala Lumpur facility is responsible for volume production of QFPs, DIPs, SOJs, TSOPs, and ceramic packages. It is currently setting up a BGA production line. It is clear that the facility makes wide use of the Motorola six sigma and 10X quality programs. [Kelly 1996]

5.3.3 Likom - Melaka

Established in 1992, Likom management has been quick to point out that the company is the "biggest OEM technopolis in Asia." Its vertically integrated facility has a current capitalization of more than $100 million and supports the following capabilities:

- design and manufacture of color computer monitors
- design and manufacture of switching power supplies, keyboards, casings and precision plastics
- design and manufacture of motherboards, PCB assemblies, multimedia and PC system unit assemblies
- tooling design for plastic injection molding
- metal stamping and a complete spray painting process for computer casings
- double and multilayer PCB fabrication

Likom's first customer in 1992 was Apple Computer. Since then, Likom's customer base has shifted to include about 75% Japanese companies. Local employees, along with 1,500 foreign workers that are approved for employment through work permits, receive hostel accommodation, meals, and transportation. Likom's highly automated plant has adopted many of its manufacturing technologies through transfer from other companies. If Likom uses the rest of the Malaysian capabilities, it will be able to deliver a product that meets all of GATT's requirements. This translates to a 3% advantage in a highly competitive market

5.3.4 Malaysian Institute of Microelectronic Systems (MIMOS)

The Malaysian Institute for Microelectronic Systems (MIMOS) is Malaysia's national center of excellence in microelectronics and information technology (IT). MIMOS started in 1985 as a branch of the Prime Minister's Department and is now a full-fledged department of the Ministry of Science, Technology, and the Environment. MIMOS invests heavily in R&D so that the domestic electronics industry will be more competitive and innovative.

MIMOS aims to promote strategic technologies for national development, to stimulate the coordinated development of an integrated electronics industry, to enhance industrial innovation and competitiveness, and to support the development of effective and efficient processes in production, manufacture, commerce, and services. MIMOS conducts research in six areas of technology: Semiconductor Technology, Design

CASE STUDIES

Methodology, Computer Technology, Product Development, Industrial Technology, and Telecommunications Technology, as described below:

Semiconductor Technology

Semiconductors are the foundation on which the Malaysian electronics industry thrives. Malaysia is the third largest producer of and the leading exporter of semiconductors in the world. It has great potential for becoming one of the world's premier semiconductor specialists. MIMOS has started to construct wafer fabrication facilities to support process and device R&D. MIMOS wants to offer expertise and facilities in silicon processing and reliability analysis. The division offers training in semiconductor technology to enhance skill availability. Its goal is to be the national center of excellence for R&D in integrated circuit silicon technology.

Design Methodology

This division provides system designers with the expertise and facilities used for IC design. It works together with the Semiconductor Technology Division to offer complete flow in IC development, from design to an operational chip. This division looks to bring the best international IC design techniques to Malaysia. It has many linkages with foreign research institutions and IC foundries to accelerate technology transfer.

Computer Technology

This division plays a major role in supporting the indigenous computer industry by developing hardware and software for domestic and international markets. The product development is mostly done in joint contracts with industry partners. The division continuously investigates technologies such as parallel processing, multimedia applications, and object-oriented techniques. This division also looks to explore hardware and software development in real-time processing, computer networking, and telecommunications.

Product Development

Through this division, MIMOS has the ability to move from basic product ideas and functional specifications into cost-effective prototypes. The goal of this division is to encourage and develop a competitive industry through product innovation, perform focused studies on industrial needs, and enhance product improvisation techniques. It routinely handles sophisticated jobs for companies involved with manufacturing of high value-added and innovative electronic products.

Industrial Technology

This division identifies focus areas for industrial automation. So far it has identified the Supervisory Control and Data Acquisition (SCADA) System, Image Processing and Pattern Recognition Technologies, and

Speech Processing and Expert Systems. MIMOS stresses the importance of applying high-technology processes in the country's industrial sectors.

Telecommunications Technology

This division's activities are directed at creating the proper infrastructure for disseminating and using information for continued national economic growth. It emphasizes the development of electronic subsystems for use in Malaysia's telecommunications and information network. The division is also studying collaborative opportunities in the development of computers and communications.

Chapter 6

GOVERNMENT AND UNIVERSITY SUPPORT OF ADVANCED TECHNOLOGY DEVELOPMENT

The governments of Singapore and Malaysia share several common traits, despite their cultural differences. Both play central roles in marshaling the resources of their cultures towards the goals of economic development and diversification. Both governments recognize that their countries (particularly Singapore) cannot depend solely on their natural resources as a source of funds to sustain the level of development desired by both government and citizenry. Both governments seek to diversify their countries' economic bases by industrialization, with a heavy emphasis on electronic manufacturing. Both provide considerable infrastructural support to investing companies: excellent facilities that capitalize on their central geographic location — roads, airports, deep-sea ports, and others; financial incentives, including tax relief, financing, and capital flow; highly trained work forces that the governments are constantly seeking to augment in terms of both size and capability; heavy investment in both basic and applied R&D at the governmental, university, and industry levels; and strong linkages between government, industries, and universities based on shared commitment to the development schema. Both governments are pledged to maintain a harmonious modus vivendi between the dominant ethnic group and the minority groups. And both have relatively stable societies with significantly low degrees of personal crime.

Perhaps the most significant trait that the Singaporean and Malaysian governments share is that their respective cultures each embrace an authoritarian notion of the role of government in society, in which the government has the right to marshal all facets of the society to achieve what it perceives as the correct purpose of that society, for the good of all. It is the citizens' duty to support their government's goals and actions. Though both countries have a form of parliamentary democracy, it is considered acceptable, even proper, that actual governmental decision-making processes and the exercise of power are highly centralized.

Solidifying this culture-based respect for government authority has been a fairly continuous rise in the standard of living for all sectors of both societies. The two governments' development plans have been very successful, and there is little reason to doubt that they will continue to be so. Furthermore, the Singaporean and Malaysian governments have actively sought to validate their citizens' support, as in Malaysia's well publicized Vision 2020 plan.

Within the context of cultural acquiescence to a strong government role in development programs, this chapter examines a number of the specific programs that the Singaporean and Malaysian governments have implemented to spur rapid economic growth in their countries in partnership with multinational corporations that provide cash and expertise in return for attractive investment incentives. In addition, this chapter examines the important development role played by the universities, also highly orchestrated by the government. A close relationship with their institutions of higher learning has been fundamental to the capability of the Singaporean and Malaysian governments to pursue their developmental courses. Local universities have been a primary source of fundamental R&D background development and of manpower training, particularly of engineers and technicians, that spur indigenous high-tech development capabilities within the two countries.

6.1 SINGAPORE

6.1.1 Government Support

In perhaps no other country in Asia does the government play a greater role in the success of the electronics industry than in Singapore. The government mixes free enterprise with guidance and grants to attract multinational corporations (MNCs) that have the cash, technology (particularly transferable technology), and capacity for growth. Lee Kuan Yew, prime minister from 1959-1990 and still the single most powerful individual in the country, introduced a series of extraordinary labor measures to attract foreign investors to the country. These include lengthening the work week; reducing the number of holidays; restricting payment of retirement bonuses, paid leave, and overtime; and exempting some promotions, transfers, firings, and work assignments from collective bargaining. These changes, though decidedly pro-business, were not achieved at the sacrifice of the worker's welfare, which is protected by other measures. [Beh 1996] Business, labor, and the government work together to advance the country's interests.

Singapore became known as the place for low-cost assembly of products for export, and these products were largely consumer electronics end-items. The industry subsequently graduated from reliance on labor-intensive production techniques to product engineering and automated assembly production techniques. Relatively recently, it has begun to perform IC design, wafer fabrication, and new product development functions.

The Singaporean government is now helping companies to expand to other nearby regions, especially the "growth triangle" Singapore forms with Indonesia and Malaysia. Singapore provides advanced technology, telecommunications, an efficiently managed transportation hub, and financing, while Indonesia and Malaysia supply land, labor (especially-non-skilled workers), water, and electric power. Working in concert with Malaysia and Indonesia, Singapore is capitalizing on its strengths to create a mutually advantageous consortium that utilizes the advantages of each of the three countries to solve the weaknesses of the others. Singapore's regional efforts extend beyond this triangle of states to include involvement in building industrial parks in China, India, and Vietnam. [Beh 1996]

Singapore has developed two government agencies that promote domestic growth in the electronics industry. They are the National Science and Technology Board (NSTB), which was established to develop Singapore into a center of excellence in selected fields of technology, and the Economic Development Board (EDB), which devises incentives to attract desired companies into the country.

National Science and Technology Board

The National Science and Technology Board (NSTB), established in 1991, is a statutory board under the Ministry of Trade and Industry that seeks to develop Singapore into a center of excellence in specific technology fields that the NSTB identifies on the basis of ability to increase the country's economic competitiveness. The NSTB promotes R&D through financial assistance schemes, coordinates the establishment of research institutions such as the Institute of Microelectronics (IME) and the GINTIC Institute of Manufacturing Technology, and attempts to develop quality R&D manpower and an R&D support infrastructure. The NSTB also builds linkages with international partners and tries to heighten public awareness of the importance of science and technology.

The NSTB, in consultation with experts from the public and private sectors, drew up the National Technology Plan (NTP) which sets the direction for developing technology in Singapore. The NTP identifies two critical targets:

1. increase the national expenditure on R&D to 2% of GDP (a minimum of 50% of expenditure must stem from the private sector)
2. increase the number of research scientists and engineers in the labor force to 40 per 10,000.

Economic Development Board

The Economic Development Board (EDB), established in 1961, has the responsibility for industrial planning and development and promotion of investments in manufacturing. Its efforts for more than three decades have resulted in the transformation of Singapore's economic landscape from that of a trading nation to that of a global manufacturing center with varied types of investment by more than 3,000 MNCs.

Through its Singapore headquarters, the EDB provides one-stop service to investors, including information and assistance in acquiring industrial land, suitable operational facilities, and skilled manpower. The EDB also offers a comprehensive range of tax and financial incentives and assists investors in obtaining financing for long-term projects. Foreign investors can use the EDB to locate customers, suppliers, subcontractors, and joint venture partners.

The EDB actively encourages the growth of local enterprise. Besides building a strong indigenous capability, it provides a sturdy base to support the development of world-class manufacturing service industry clusters. The Board focuses as well on helping local companies identify opportunities and start operations in the region.

One of the EDB's key strategies is to build depth in technology-intensive manufacturing and higher-value-added activities where Singapore has a competitive edge and comparative advantage. It actively promotes Singapore's total business capabilities so that investors can set up the entire range of their business activities in one location.

Singapore's regionalization thrust (vis-à-vis Malaysia and Singapore within its immediate geographical locale, the ASEAN members in the slightly wider sphere, and beyond that, countries such as China and India) creates an external wing of its economy and further enhances its attractiveness as a business location. The EDB provides facilitation assistance to companies that use Singapore as a base to develop, manage, and invest in the entire region. The Board continues to strengthen the information flow between the private and public sectors to identify potential projects and partnership opportunities.

6.1.2 University Support

As does any thriving industry, the electronics sector needs a competent work force that can adapt quickly to changing technological trends. This includes both well-trained assembly workers and proficient engineers. Singapore's educational institutions prepare the manpower for the demanding electronics industry with a focused academic training program and by allowing students access to high-technology R&D projects. Singapore's two universities, the National University of Singapore and Nanyang Technological University, each house high-technology research institutions in which students participate in ongoing projects.

National University of Singapore

The National University of Singapore (NUS) was formally established in 1980 with the merger of the University of Singapore and Nanyang

University. The university has four specialized advanced training and research institutions, including the Institute of Microelectronics (IME). In addition, it has several research centers within academic departments, two of which are involved with electronics: the Center for Integrated Circuit Failure Analysis and Reliability and the Center for Optoelectronics.

The Institute of Microelectronics (IME) is a national industrial R&D laboratory whose mission is to help increase the value-added factor of Singapore's electronics products. IME helps determine which microelectronics R&D projects should be focused on. Five areas of current interest to the IME are VLSI circuit design and testing; microelectronics systems applications and mobile communications; microelectronics process technology; failure analysis and reliability testing; and advanced packaging development support. IME also supports and forms partnerships with local electronics companies and helps train their skilled personnel. Its staff totals approximately 160; its funding comes primarily from the NSTB.

The Center for Integrated Circuit Failure Analysis and Reliability (CICFAR) conducts research to investigate, understand, and develop techniques for the diagnosis of problems related to IC failure analysis and reliability testing. The center provides specialized consultation services to industry and organizes industrially relevant short courses in collaboration with industries and other research centers. It has collaborative research projects with industrial giants such as AT&T, Singapore's Institute of Microelectronics (IME), National Semiconductor, Siemens, and Texas Instruments.

CICFAR is well equipped for the failure analysis and reliability testing of ICs and optoelectronic devices. It has a wide variety of failure analysis techniques, including scanning electron microscopy (SEM), SEM voltage contrast (static and dynamic), SEM electron beam induced current (EBIC), SEM cathodoluminescence (CL), panchromatic and monochromatic imaging, photoemission microscopy, tunneling current microscopy (TCM), scanning acoustic microscopy (SAM), scanning tunneling microscopy (STM), atomic force microscopy (AFM), magnetic force microscopy (MFM), charge pumping and floating gate measurement techniques for hot-carrier reliability studies of MOSFETs, liquid crystal hot-spot detection and voltage contrast, and on-die stress measurement of IC packages.

Five projects form the backbone of CICFAR's research plan for the next five years: (1) Failure analysis of the backside regions of ICs; (2) advanced SEM techniques for IC inspection; (3) spectroscopic photoemission studies of semiconductor devices; (4) electron beam testing of ICs; and (5) hot-carrier reliability studies of submicron MOS devices.

The Center for Optoelectronics (COE) was established to achieve excellence in research in the emerging optoelectronics technologies and to catalyze the development of the optoelectronics industry sector in Singapore. The COE obtains knowledge in the field by conducting research projects and through participation in international optoelectronics forums. The COE actively participates in technology transfer to domestic industries by collaborative research projects, courses, and seminars. It also seeks to develop efficient manpower in the field. Some of its current research projects include fabrication of semiconductor lasers, study of the fundamental processes of semiconductors, and semiconductor device modeling.

Nanyang Technological University

Nanyang Technological University (NTU) was founded by Parliament in 1991. Research and development form an integral part of the activities at NTU. The university encourages staff to do R&D work as well as to provide consultation to industry. Funds are provided for staff to carry out approved research projects. One of the research institutions involved with the electronics industry is the GINTIC Institute for Manufacturing Technology. To encourage local and locally-based companies to carry out R&D activities on campus, NTU has an Innovation Center that acts as a miniature science park. To date, 20 firms have established units at the Innovation Center for research and development work.

The GINTIC Institute of Manufacturing Technology was formed as a national research institution funded by the NSTB. GINTIC performs applied R&D in identified processes and develops technologies designed to help develop Singapore's manufacturing industry. As the electronics industry is the core of Singapore's manufacturing base, it benefits heavily from GINTIC's research. GINTIC helps the locally-based manufacturing community to be more competitive by transferring various technologies as appropriate. It offers special R&D services in printed circuit board assembly and surface mount technology.

6.1.3 University and Government Relationship

In Singapore (and also in Malaysia), the relationship between the government and its educational system, particularly in higher education, is profound and multifaceted. The universities are solidly supportive of the government, its programs, and its direction of development. The Government of Singapore, through the NSTB and EDB, strongly affects the way that the universities and the entire educational system function. By mixing guidance with grants and attraction of MNCs and other companies to the universities, the government emphasizes the universities' role in its overall development program. Since Singapore is increasingly dependent on the value-added edge that a highly skilled work force brings, the role that these educational institutions play in creating this work force is becoming increasingly important. Industry has, partly through government action, of course, realized the fundamental role that these educational institutions play. Along with the government, industry is supporting the universities and aiding in their research and development programs and education of upcoming scientists, engineers, and skilled workers. The government-

6.2 MALAYSIA

6.2.1 Government Support

Up until the early 1990s, foreign investors were wary of entering the Malaysian market. The government had been reluctant to provide incentives to investors, due largely to the fact that the government wanted the indigenous population to better the economy on its own. Comparing the economy in 1990 to that in 1969, Malaysians have had some remarkable achievements in this regard. However, in order to maintain Malaysia's high rate of economic growth, the Malaysian government has realized, particularly in the last five years, that it must bring leading-edge technology to its shores, and that foreign direct investment (FDI), is an effective tool to achieve this end. With this realization has come a willingness to provide incentives to foreign investors, to pursue an assortment of other initiatives that strengthen the overall investment infrastructure, to invest heavily in economic management vehicles through such agencies as the Malaysian Institute for Microelectronic Systems (MIMOS), and to continue to create and religiously follow master development plans — in today's case Vision 2020 — as has been done since 1969.

Financial Incentives

Malaysia has now developed one of East Asia's most comprehensive fiscal packages to encourage both internal and external investment in high-technology R&D and establishment of high-tech industries. The Malaysian government grants special Pioneer Status and/or Investment Tax Allowances to these ends, including those listed below:

- A full income tax exemption (Pioneer Status) or Investment Tax Allowance of 60% on qualifying capital expenditure for five years is granted for high-tech projects. The allowance can be offset against 100% of statutory income for each assessment year.
- A full income tax exemption (Pioneer Status) or Investment Tax Allowance of 100% on qualifying capital expenditures for ten years is granted to contract R&D companies. The allowance is granted at statutory income level, and abatement for each assessment year is limited to 70% of statutory income.
- An Investment Tax Allowance of 50% on qualifying capital R&D expenditures for five years is granted to companies carrying out in-house research. The allowance is granted at the statutory income level, and abatement is limited to 70% of statutory income.
- Companies that establish technical or vocational institutions are eligible for a 100% Investment Tax Allowance for ten years. Existing technical

or vocational training institutions that upgrade equipment or expand their capacity are eligible for the same incentives. These institutions are also eligible for import duty exemption on materials, machinery, and equipment used for training.

- A double deduction is granted for non-capital expenditure on either in-house or contracted research projects.
- A double deduction is granted for non-capital expenditures incurred by companies that use services provided by approved research companies or institutions.
- A double deduction is allowed for cash contributions to approved research institutions.
- Capital allowances are offered for plant and machinery used for approved research.
- Companies can claim exemption from import duty, excise duty, and sales tax on equipment, machinery, raw materials, and samples used for approved research projects.
- Industrial building allowances are given for buildings used for approved research projects.

Other Initiatives

In addition to financial packages, the Malaysian government has established and maintained organizations, funds, and technology parks to aid the technological research and development process. These all have received significant government support and funding. Some of these measures include the following:

- Establishment of the Intensification of Research in Priority Areas (IRPA) Fund in 1988 to provide financing for research projects that have commercial potential. This fund encourages government-owned R&D agencies and universities to have greater linkages with the private sector. The microelectronics industry has benefited most from this fund.
- Establishment of the Malaysian Technology Park in 1988, which has assisted in developing indigenous technologies and commercialization of R&D findings. Its main role is to support industrial high-technology entrepreneurship. In addition, it promotes industrial competitiveness, encourages reverse engineering, and accelerates technology. The Malaysian Technology Park provides a vital link between industry, government, R&D institutions, and universities.
- Launching of the Kulim Hi-Tech Industrial Park (KHTP) to aid in the development of high-technology industry. Its objectives are to be a center of excellence for high-technology industries and R&D and for technology training.

- Creation of the Industrial Technical Assistance Fund (ITAF) in 1990 to provide financial assistance to small- and medium-scale industries. The assistance is in the form of matching grants for consulting studies, product development and design, quality and productivity improvements, and market development.
- Establishment of the Malaysian Technology Development Corporation (MTDC) in 1992, a government-industry joint venture that focuses on commercializing local research findings, introducing strategic technologies to the country, and manufacturing products widely used as industrial inputs. It also acts as a catalyst to development of venture capital and as a center for growing technology-based companies.
- Establishment in 1993 of the Malaysian Industry Government Group for High Technology (MIGHT), a government-private sector group initiative to exploit research and technology for new business opportunities.
- Government relaxation of conditions for employment of skilled expatriates and foreign workers who are involved in R&D activities.
- Pursuit of a number of resource development measures to increase the number of skilled workers. This includes expanding and upgrading existing industrial training institutions and developing new polytechnical and industrial training institutions.

Free Trade Zones (Science Parks)

As previously mentioned, the Malaysian government has assisted in establishing centers for the promotion of high-technology industries. Called Free Trade Zones, these industrial parks operate under the tenet that a group of industries working in a common environment can accomplish more than if they remain isolated. The government's strategies to accomplish this include geographically concentrating public and private research institutions, promoting hybrid technologies, upgrading local university laboratories, establishing technology centers, and funding both joint and individual R&D projects. As noted above, two of the more well-established examples of these industrial parks in Malaysia are the Technology Park of Malaysia and the Kulim Hi-Tech Industrial Park.

The Technology Park of Malaysia (TPM), the country's first science park, opened in 1988. It was established by the Ministry of Science, Technology and the Environment to stimulate indigenous technology development. It is located in Kuala Lumpur near five universities and nine research institutions. TPM integrates state-of-the-art facilities, services, and conveniences. The proximity of major universities promotes a symbiotic relationship between TPM's tenants and academia. The infrastructural facilities include board rooms and conference rooms. The park's laboratories for precision engineering and microelectronics will be equipped with advanced equipment and tools.

TPM is functionally divided into four physical components: the Innovation House, the Incubator Center, the Enterprise House, and R&D lots. These divisions enable TPM to support the growth and competitiveness of companies through the different stages of product development to manufacturing. The Innovation House is comprised of small modular units where companies can set up a research and development base. The Incubator Center is suitable for small-technology-based startup companies. The Enterprise Houses are self-contained areas suitable for medium-sized businesses related to industrial production of high technology. Each R&D lot is spread over an area from one to five acres that can be leased on a long-term basis. The lots cater to large independent tenant companies that require special building arrangements such as R&D innovation and limited production runs.

TPM also provides marketing services for tenant companies. It actively participates in trade fairs that help tenant companies to establish global linkages with potential partners and customers. Through its Information Center, TPM provides easy access to patent information, a registration office, a computerized data bank, and a wide range of scientific and technical journals.

The Kulim Hi-Tech Park (KHTP) is an industrial park developed by the central government in conjunction with the state of Kedah. The 3,600 acre park was designed within a detailed environmental framework to blend high-technology manufacturing, training programs, and R&D activities with a healthy contemporary lifestyle.

KHTP induces companies to locate in the park by making relocation costs competitive and by customizing industrial sites to fit specific needs. The Park is located near Penang, a major transportation hub in Malaysia. The Technocenter and the Information Technology Center located within the Park provide industries with scientific and technical support. The KHTP Local Authority has a "one-stop center" for development approval.

KHTP's Technocenter provides R&D support for industries in the park and surrounding areas. Technocenter facilities include R&D areas, incubation facilities, and human resources training facilities. A study of Technocenter management and planning has been completed by the Japan International Cooperation Agency (JICA).

KHTP's Information Technology Center (ITC) is the first of its kind in Malaysia. Its main role is to execute comprehensive research in information technology. The facilities include CIM Workstations, CAD and/or CAM facilities, a PC training lab, demonstration and/or testing CIM facilities, and rentable office spaces. It is surrounded by a number of lots where software development firms will be invited to locate themselves, making the area an information technology core called the KHTP Software Park.

In addition to establishing industries, KHTP in cooperation with University Teknologi Malaysia operates an institution of higher learning at the park called the Standard and Industries Research Institute of Malaysia

MALAYSIA 81

(SIRIM). Additionally, the Malaysian Institute of Microelectronics (MIMOS) is starting a microelectronics research center at the park. KHTP wants to set-up several higher learning institutions so that it can become the national center of excellence for high-tech training.

6.2.2 University Support

The Malaysian government understands well the role that educational institutions must play in producing engineers and technicians who can support Malaysia's development needs. From the earliest school years on, particularly as laid out in *Vision 2020*, the government seeks to produce school graduates who not only have the necessary skills, but also the "proper attitude" to contribute to the development of the country. Government leaders feel strongly that without proper academic training of the populace, Malaysia will not have the competitive intellectual strength to propel the country into the 21st century.

Along these lines, universities have played a major role in supporting the growth of the electronics industry. For example, Malaysian universities have invited industries to establish independent and joint R&D facilities on their campuses. These actions are, of course, heavily supported by government financial, logistical, and political support. Two of the larger universities that participate in such programs are Universiti Science Malaysia and University Teknologi Malaysia.

Universiti Science Malaysia

Universiti Science Malaysia (USM) is a science and engineering institution that offers industrial companies facilities and expertise on campus via contract research, consulting, routine tests, courses, and training. The university also allows companies to establish their own R&D groups on campus.

USM stresses R&D, design, engineering, and prototype activities leading to manufacture. Research at USM is geared towards solving problems relevant to industry so that productivity and efficiencies are improved. Much of this achievement is due to the Industrial Consultancy Unit, now known as the Innovation and Consultancy Center (ICC), founded at USM in 1981.

ICC has gained a reputation as being a blueprint program for those wanting to set up a successful university-industry relationship. It aids in promoting R&D in addition to acting as a resource center for expertise from university faculty and staff. Presently more than 700 organizations have used the ICC's services, and 12 companies have actually set up shop on the campus.

University Teknologi Malaysia

University Teknologi Malaysia (UTM) is the largest technological center for education in Malaysia. The school employs more than 1,000 scientists and technology specialists who, aside from being academicians, are dynamically involved in research, publishing, and consulting activities. All 10,000 students at the school take courses in science and technology.

Most of the research and consulting activities are coordinated by the Research and Consultancy Unit (RCU) at UTM. The objectives of RCU are to enhance the technological development of the nation through research and consulting and to facilitate and coordinate research and consulting activities undertaken by UTM. RCU draws on the expertise of the many faculty members at UTM.

In order to improve the management of consulting activities, RCU formed a company called Uni-Technologies in 1992, in agreement with the UTM Council. In line with the formation of the company, an innovation center was developed to operate as a center for commercial activities.

6.2.3 Government and University Relationship

"Mainly we need people who are skilled...that is why we are introducing computer science...(and) ...are making people familiar with machines, engines and electronic parts." (*Vision 2020*)

"We want to actively change their value system. We are setting up centers where we will take people and tell them why this value system is bad and why this is good and why you should practice this value system and not that. We have devoted RM$ 100 million to this process." (*Vision 2020*)

As indicated by the above quotes from *Vision 2020*, the basic planning document put out by the Government of Malaysia, the government is committed to adjusting and reforming its society through education in order to pursue its development goals, which are heavily biased towards accelerating development in the electronics industry. This commitment informs the strong interrelationship between the government and Malaysia's educational system. As the Vision 2020 comments indicate, the government expects that change in society will depend heavily upon inculcation in the people of specific values and skills by the educational institutions. These institutions will through government intervention and direction, financing, monitoring of actions, and other support, be strongly influenced by the government's goals and plans, and will, in turn, influence to a major degree the formation of the work attitudes, job skills, and future employment and career opportunities of the Malaysian people.

In summation, the government's role in education is increasing and becoming more intensive as the government uses the educational institutions to inculcate the values and skills it feels are necessary to serve as the basis for the country's future development.

Malaysian universities and other educational institutions clearly benefit from this increased attention and involvement of the government. They receive substantial government funding; in addition, through government encouragement of industrial participation, they receive not only additional

SUMMARY

sources of funding but also material and equipment and opportunities for research and training for both students and faculty that might not be otherwise available. The obvious drawback for the universities and other educational institutions is that their programs, mode of education, and direction will now be greatly, and perhaps increasingly, restricted.

6.3 SUMMARY

In summary, the relationship within Malaysia and Singapore of the governments to the universities and educational systems is very intense. Both governments are heavily involved in not only how these educational institutions are run but in what they are teaching. They influence the universities through direct governmental funding, directives to create special programs, aid in generating funding and other support from industry; cooperative programs between university and government agencies; and technical support. Both governments see the role of the universities and other educational institutions as being fundamental to the success of their national development plans. They are seen as being essential to the generation and continuation of R&D efforts at all levels of the process and to the production of a highly motivated and skilled work force. In Malaysia's and Singapore's game plans for their continued and rapid economic development, the government, the educational institutions, and industry all play vital and interdependent roles.

Chapter 7

CONCLUSIONS

Singapore and Malaysia are remarkable among the countries of Southeast Asia in being economically sound and increasingly competitive internationally; in having good trade balances, good capital flow, and diversified economic bases; and in not being heavily indebted to the World Bank or IMF. A major portion of their trade expansion is occurring in the electronics manufacturing sector.

Singapore and Malaysia did not achieve their enviable economic strengths overnight or without social costs. Since achieving independence from Great Britain after World War II, both countries have undergone various sociopolitical upheavals that have had a major impact upon the roles of government within both societies and upon the people's roles vis à vis their governments. For example, the 1969 race riots in Malaysia, the subsequent quelling of the riots and the two-year state-of-emergency, forever altered the nature of Malaysian society and the manner of government that drove it. The prime goal of the Malaysian government since this period has been to foster economic diversification and industrial development as the basis of a steady march towards modernization; Malaysian society, which had been basically rural, has been guided along a course of modernization as defined by the government, which definition does not include preserving the rural-based society of the past.

Singapore has undergone sociopolitical changes as well. As the governmental decision-making process became increasingly centralized after Singapore's separation from the Malaysian Confederation in 1965, the government assumed an increasingly dictatorial role within the society. As the other "mini-dragons" of Asia were doing at approximately the same time, the government of Singapore began to focus its efforts and marshal the society's resources towards development.

Though progressing at different rates along the path of modernization, and despite racial, linguistic, economic and other differences, Singapore and Malaysia share a number of societal-cultural factors that create a similar environment for development:

- Both are basically single-party countries, and the primary culture of each (Chinese in Singapore, Malay in Malaysia) dominates the country through these parties. Each country's culture to a great degree grants the state the right, through a relatively centralized decision-making process, to marshal resources to meet the goals determined by the dominant group.
- Both governments are committed to an intense program of long- and short-term economic development based primarily on an export orientation. This developmental course includes a program of economic diversification where commodities account for a decreasing share of GNP and electronics account for an increasing share, particularly in Singapore, which has the capacity for advanced electronics manufacturing, including end-assembly operations. For example, Singapore Technologies will be the exclusive manufacturer of a new line of high-end Apple Computer monitors under an agreement that will earn it an estimated $150 million a year starting in 1997.
- Both governments are strongly focused on the goal of forging their own countries into major East Asian economic powers. It is perhaps ironic, considering their initially difficult post-colonial relations, that today the two countries are working very closely together: In 1995, 450,000 Malaysian workers had jobs in Singapore, and Singapore is transferring technology to Malaysia as the two countries work together on joint projects. However, Singapore is establishing a certain hegemony over the immediate region as the major economic powerhouse with its development of industrial parks, cooperative agreements and projects.

This chapter analyzes and summarizes these developments as they have been discussed in the body of this work. It is broken down, by country, into three general sections: the country's current state of development as reflected in current structural characteristics that affect its future development; its future developmental plans; and the role of industry in the country's plans, with an emphasis on the role of the electronics sector.

7.1 SINGAPORE

7.1.1 Current State of Development

Singapore's current state of development provides it with a solid basis for its planned development into the future. Singapore's plans have been remarkably consistent and forward-looking. The present state of the economy both embodies the realization of previous plans and acts as a stepping stone for the realization of ongoing plans. Currently, plan "E 2000," which is Singapore's overall plan for future development and roughly equivalent to Malaysia's "Vision 2020" plan, is the driving force behind development activities, but it was preceded by other just as far-reaching goals and plans.

Singapore's infrastructure is excellent. Geographically located at one of East Asia's historical hubs, it has long served as a focal point for commerce, development, and communication in this area of the world. It has continually updated and developed its geographic advantages to meet the needs of the modern world so that today it has some of the best seaports and airport facilities of the world, designed to handle the most sophisticated of cargo and transportation requirements. In addition to these, it has a modern network of roads and an efficient mass transit system comprised of bus, train, and light rail. Together, these systems service all of Singapore's manufacturing, business, shopping, and residential areas, creating an interlocking infrastructure that easily supports not only the current level of transit needs, but also provides a solid basis for further development.

Complementing Singapore's transportation infrastructure is its extensive telecommunications system that is completely digital and utilizes the most modern technology available. The system is constantly being upgraded to provide users with a competitive edge, indicative of Singapore's determination to provide a support infrastructure second to none for the country's development.

As regards power and water, the country has constant but increasingly expensive supplies of both. These resources are developed both internally and in conjunction with Malaysia and Indonesia. The regional approach to solving problems of this type is typical of Singapore's approach to its relationships with its neighbors; old antagonisms have been to a great degree buried under requirements to meet industrial needs in as cost-efficient and effective a manner as possible. Employment of a large number of Malaysians, joint development ventures with Malaysia, and a degree of technology transfer to that country belie the animosities, particularly racial, that contributed to Singapore's separation from the Federation of Malaysia in 1965 and to Malaysia's 1969 riots that were aimed largely at its ethnic Chinese sector. Pursuit of business has become the paramount concern.

In Singapore, the government-industry-university interrelationship in support of the country's economic development is especially strong. As the government seeks to increase the country's domestic research efforts, it is working with the universities to increase support for research as well as to develop the research results. The government involves industry in this effort by such actions as developing commercial parks and encouraging industry cooperation in university R&D efforts. The same pattern holds for government-university-industry cooperation to develop the highly skilled work force required by Singapore's development plans.

The relationship between Singapore's government and its educational system, particularly its institutions of higher learning, is profound. The educational system strongly supports the government, its programs, and its development direction; likewise, the government through the NSTB and EDB strongly affects the way that the educational system functions. By mixing guidance with grants and by attracting multinational and other companies to the universities, the government gives great emphasis to the role of the universities in its overall development program. Since Singapore is increasingly dependent on the value-added edge that a highly skilled work force brings, the role that Singapore's educational institutions play in creating this work force is becoming increasingly important.

Industries in Singapore play a strong role in the universities, primarily due to government encouragement. Industries provide monetary support to the universities, establish R&D facilities (in contrast to the past, research is increasingly being emphasized over development), help to direct training of students, and in league with the government, help to increase the number and quality of technologically skilled workers. This role obviously works to the benefit of industry by providing aid in R&D for new products and processes and by increasing the pool of skilled labor.

In cooperation with the industrial and educational sectors, the Singapore government is preparing its work force of the future by the following kinds of actions:

- improving the overall education level of its work force
- offering technical education incentives concentrating on targeted areas
- building technical institutes to provide concentrated training
- sponsoring technical seminars and utilizing foreign experts to increase dissemination of new knowledge and materials and increase effectiveness of spot training
- encouraging MNCs to provide more worker training to its work force, to increase technology transfer
- increasing the numbers of Singaporeans educated in foreign universities and returning home with their new skills and knowledge

Finally, the effectiveness of its capital development system is another structural element that contributes significantly to Singapore's rapid economic development. The government uses a number of incentives to attract capital in any form, for example, investment allowances, expansion incentives, exporting of service, approved foreign loan schemes, approved royalties, overseas enterprise incentives, venture capital incentives, operational headquarters incentives, business headquarters, double reduction for R&D expenses, double deduction for overseas investment development expenditures, research incentive schemes, coinvestment incentives, and so forth. Singapore is also working on reducing its income tax rate for corporations. With these types of incentives, manufacturing investments in 1994 reached $4.1 billion, an increase of 49% over 1993. By the end of 1994, official foreign reserves were $85 billion. An unofficial, tongue-in-cheek comment sometimes heard in Singapore is that the country has more reserves than methods to spend them — Singapore has a current yearly budget surplus of $10 billion.

Singapore, like many other Asian countries is a major investment area for many other countries — chief among them, Japan and United States. Singapore has gone to great lengths to deregulate its investment and capital flow procedures, making direct investment there very attractive. In its new 5-year plan that started in 1996, Singapore seeks to further increase both its inflow of capital through FDI and its export of capital through an intensive regional development plan that will help it to maximize its own internal resource capabilities. By such means as moving lower-end technology production to countries like Malaysia and Indonesia, it will build stronger economic ties with those countries and also more effectively use its own labor force for technologically sophisticated end-work, providing a higher dollar return on investment. This will also help Singapore to economically dominate the region and to compensate for such shortcomings as minimal natural resources and relatively small local markets, labor force, and geographic area.

7.1.2 Role of Industry - Specifically Electronics

Throughout Singapore's economic development over the past three decades, electronics has played a major role and continues increasingly to do so. Output has been skyrocketing; in 1994, output reached $32.9 billion, accounting for almost 50% of total manufacturing output. This was a 24% rise over the previous year. Singapore firms Creative Technology and Aztech Technology together produced almost 90% of the world's sound cards. The supporting industries that have developed in Singapore in order to provide a wide range of precision engineered parts, components, and services to support the manufacture of disk drives, computers, color televisions, VCRs, instrumentation, and other precision engineered products contributed over $800 million in goods and services to the economy.

Singapore's electronic data processing sector is the clear leader in terms of electronics production. This sector produced over $19 billion in goods in 1995, accounting for over 50% of the country's total electronics production. The domestic computer manufacturers are getting the world's attention with their competitive costs and highly reliable IBM compatibles. Disk drive exports alone accounted for 39% of total computer equipment exports in 1995.

Additionally, production of consumer electronics over the last four years has shown a great deal of stability in terms of employment and value added. Production had a compounded average annual growth of 10%. Output was $3.9 billion in 1992, $4.1 billion in 1993, $5 billion in 1994, and $5.2 billion in 1995.

The data storage sector leads Singapore's electronics industry in terms of monetary value of output. From 1992 to 1994, production in the sector increased from $5.7 billion to $7.3 billion, an annual average compound growth rate of 14%.

Office automation equipment has been another booming growth sector in Singapore's electronics industry. From 1992-1994, output in this sector increased from $1.53 billion to $3.33 billion. Compared to this large increase, the averaged employment rate has only increased 6%, indicating a large value-added production rate where production efficiency and innovation levels have increased dramatically.

These points clearly illustrate the critical role that electronics has played and continues to play in Singapore's economic growth, and it has done so with phenomenal efficiency. For example, TI's Singapore operations enjoyed 99.5% on-time delivery rate for the first quarter of 1996.

Within the E 2000 cluster plan, electronics will continue to play the leading role among the six targeted market clusters: it is planned to contribute 51% of output. The electronics cluster consists of consumer electronics, communications, semiconductors, electrical supporting industries, computers, and mass storage and peripherals. Related areas include chemicals; application intelligent software; multimedia, PDA, DSS, and set box tops; and general supporting. As an indicator of the future success of the new plan, it is worth noting that in 1995 alone, the electronics cluster's output was approximately $42 billion. From 1994 to 1995, data storage devices grew almost 20% ($8.1 billion to $9.6 billion); semiconductors grew 53% ($5.7 billion to $8.7 billion); computers grew 10% ($7.4 billion to $8.2 billion); communications grew 37% ($1.9 billion to $2.6 billion); office automation grew 14% ($2.1 billion to $2.4 billion); passive CPTs and PCBs grew 7% ($1.4 billion to $1.5 billion); and display devices grew 33% ($700 million to $930 million). In the less active fields, consumer electronics dropped 4% ($5.3 billion to $5.1 billion) and contract manufacturing and others retained a level output ($2.4 billion).

Singapore's E 2000 plan seeks to take the 1996 base of six fabrication facilities (fabs) and within ten years increase them to 20 fabs. It is important to note here that all current capabilities are required for future growth, and the plan addresses them as such.

Singapore's readiness to grasp the future of electronics technology is underlined in the increasing number of research scientists and engineers (RSEs) being trained. In 1993, RSEs numbered well over 2,000, and accounted for one-third of the total number of RSEs in manufacturing. In addition, Singapore's annual expenditure per RSE exceeds Taiwan's and Korea's by $116,000.

Further examples of Singapore's drive to master electronics technology are the planned creation of additional "Centers of Excellence" and further development of existing centers, as well as attraction of more MNC investment. For example, TECH, a consortium of TI, Canon, Hewlett-Packard, and the Economic Development Board, will construct a 64-Mbit DRAM fabrication facility. It will cost $450 million and be TECH's second. It will raise the company's monthly wafer output to 25,000 and is expected to generate $600 million in revenues by 1999.

There are three major points to stress about Singapore's future course of growth and development, especially with regard to its electronics industry:

1. Plan E 2000 emphasizes moving Singapore to more advanced production capabilities, placing strong emphasis on technology and innovation. With

the country's inherent labor shortage, this is the only way that it will be able to achieve its growth goals.
2. The plan emphasizes increasing cooperation between government, industry, and academia. Singapore has the widest range of programs in East Asia to support development of targeted industries, such as semiconductors, communications, displays, and data storage. Fundamental to this approach are close coordination and mutual support of these three main sectors of the society.

Through active membership in ASEAN, Singapore and the other member countries both individually and as a group, both domestically and regionally, are investing heavily to improve their core competencies in all links of the supply chain. Singapore invested $28 billion in 1993 in other countries of the region. As it demonstrated in its response to the world recession of the early mid-1980s, Singapore has successfully met major challenges to its efforts and continued to press forward. Its very long term goal of surpassing the United States' per capita GNP by the year 2030, may indeed be feasible, considering its almost consistent high growth rate and low inflation rate (as of 1996, GDP growth greater than 8% per year with inflation less than 4%). This is particularly true with its strategy to increasingly create higher value-added products, develop greater R&D activities, create a better trained work force, and further develop its regional leadership and role as gateway to that area for MNCs and other countries to channel investment, transfer technology, and so forth.

7.1.3 Future Plans

The goals of actions of the Singapore government that provide a basis and start for its next set of goals are encapsulated in plan E 2000. As Chairman of Singapore's EDB Philip Yeo has stated, the national vision of Singapore in plan E 2000, is simply to "provide our youth and future generation with meaningful jobs, high standard of living, and a quality of life second to none." Current conditions provide a solid basis upon which to grow and develop by maximizing capabilities and eliminating or minimizing weaknesses, as in the following plans:

- improve education
- increase the number of technical institutes
- further develop trade-sale and/or marketing
- upgrade telecommunications (fiber, wireless, etc.)
- improve mass transit to achieve the state of the art
- increase the presence of foreign technical experts
- import laborers from such countries as Bangladesh, Indonesia, Malaysia, and the Philippines
- form technical associations and technical consortia (e.g., the ball grid array consortium)

On this base, Singapore has developed a very clear, six-pronged strategy of development (see also Chapter 6):

1. Develop strategic industry alliances between technology-intensive local companies, MNCs, and support industries to drive further growth.
2. Make Singapore into a regional business hub by inducing MNCs (with the kinds of relocation and investment incentives discussed above) to make Singapore their operational regional headquarters and/or form partnerships with Singaporean companies. Singapore has already attracted 50 MNCs to the region.
3. Promote regional development, forging strong links to neighboring countries by building partnerships with regional companies, enhancing the bond with developed countries and MNCs, and expediting government-initiated regional development plans. The ultimate goal is to recreate Singapore's previous position as the gateway to the region.
4. Develop new business councils to seek out international partners, and promote and simplify cooperation in order to profit from opportunities Singapore cannot exploit alone.
5. Support the development of secondary "S"-curve companies into MNCs through development of strategic plans, new products, and bigger markets. Many incentives currently in use focus on these companies, such as by granting them "pioneer" or "post-pioneer" status.
6. Sell itself through collective marketing as a country where government, business, and academia work together to satisfy common goals. For example, the government, working with and through industry and academia, is training specialist manpower in precision engineering, factory automation, IC design, and surface mount technology, and it is establishing national centers of excellence such as the Institute of Microelectronics, and GINTIC's Institute of Manufacturing Technology to focus on applied R&D.

The comprehensive plan of government leaders is to continually improve Singapore's GDP through technology investments targeted at emerging markets in electronics, biotechnology, telecommunication, information technology, wireless communications, and multimedia-PDA-set top boxes. With the EDB and NSTB to organize and develop the goal of providing Singapore's citizens with a first-rate standard of living, government leaders seek to maintain manufacturing at above a 25% share of total GDP. The EDB through its growth strategy works with industry to develop technology road maps that support targeted market "clusters," which are defined based on their core competency and market potential. In E 2000 the cluster groups are aerospace, general supporting, heavy engineering, electronics, chemicals, and precision engineering. In this new plan, the EDB has a cluster development fund of $1 billion to facilitate rapid expansion in these areas.

7.2 MALAYSIA

7.2.1 Current State of Development

Malaysia is an Asian "mini-dragon" in the making. Currently involved primarily in low-cost manufacturing, it is making rapid strides along the same path of development that Singapore and the other mini-dragons have followed as they, in turn, have followed the "dragon" country Japan. Malaysia continues to make remarkable strides in its development, exhibiting a rapid growth pattern that is fairly typical of the newly industrialized countries such as its neighbor Singapore. While not as fully developed as Singapore, it is nonetheless highly developed for a Third World country, is able to accommodate its current needs, and is actively planning for future development.

Like Singapore, Malaysia has an excellent geographic location, partly exploited in the past by the British as a gateway to Southeast Asia. It has deep-water ports and modern airport facilities — not of as consistently high a quality as Singapore's, but of sufficient quality to support its growth. Their upgrade is part of the government's development program. Also, despite Malaysia's large geographic area, its roads and transportation system adequately provide ready access to commercial areas, schools, and residential areas. With global reach through its ports, airports, internal transportation arrangements, and telecommunications facilities, and with relatively constant water and power supplies, the country has the basic infrastructure in place upon which to build further in order to meet developmental goals.

Malaysia's commercial and industrial areas are basically suburban, more dispersed as compared to Singapore's concentrated arrangement. For example, Malaysia has created eleven free trade zones (FTZs) to act as incentives to development in specific geographic areas located strategically throughout the country. Within these areas, companies are subject to minimum customs formalities and are exempted from import duties on raw materials, machinery, and component parts. For electronics, these areas are Batu Berendam (Malacca), Ulu Klang (Kuala Lumpur, or KL), Bayan Lepas (Penang), Prai (Penang), Technology Park (KL), Kulim Hi-Tech Park (Kedah), Shah Alam (Selangor), Jahore Technology Park (Jahore), Sungei Way (Selangor), and Subang Hi-Tech Park (Selangor).

Like its transportation linkages, Malaysia's telecommunications system is also not as sophisticated as Singapore's but more then adequately joins the various facilities together. It is another area that the Malaysian government is focusing on upgrading. Along with developing these systems, Malaysia is working with Singapore and Indonesia to further develop and refine the supplies of water and power of the region. In the area of natural resources, Malaysia has an edge relative to Singapore because of its petroleum reserves, which are sufficient to allow some exportation after meeting internal consumption needs.

A major infrastructural element supporting Malaysia's development is the close relationship between its industrial, academic, and government sectors. This close relationship, similar to that of Singapore, Taiwan, and South Korea, is fundamental to Malaysia's development style. All of the country's series of development plans, including the currently planned 7th and the "vision" that was laid out in *Vision 2020* (1996), place increasing emphasis on this relationship. The basis for this emphasis is twofold: (1) to more rapidly increase the capability of the manufacturing sector to produce higher value added products, and (2) to expedite the development of a higher grade of worker, imbued with a strong work ethic and with company loyalty, who is capable of meeting the needs of the industries as they increase their manufacturing sophistication. The Malaysian government has pursued, and is pursuing, a three-step plan:

1. create preemployment schools to imbue the employees with a greater sense of company and national loyalty and with a stronger work ethic
2. create and further develop hi-tech parks that concentrate industrial production, R&D, and academic education
3. amplify technical training and R&D programs at the universities and firmly emphasize getting industry involved in these programs, in order to further R&D efforts and to provide a higher quality of worker in order to move away from labor-intensive industries towards technology-intensive industries.

Another infrastructural element strongly supporting Malaysia's economic growth is its capital development program. A clear example of the country's investment-friendly atmosphere is its provision to allow in some cases 100% foreign ownership of particular industries. Another example is the eleven FTZs where companies receive a wide variety of tax advantages. A third example is the seed capital that MTDC provides to companies interested in investing in Malaysia. MTDC also provides capital to local companies involved in technology-based operations to help them enhance their competitiveness. MTDC's capital development objectives are (1) to develop indigenous technology, (2) to facilitate industry's absorption of technology developed by universities and other institutions, (3) to transfer technology to local companies by gaining access to foreign technologies, and (4) to develop venture capital activities in Malaysia.

To achieve its development objectives, the government of Malaysia offers a wide variety of attractive tax incentives that range from pioneer status (partial exemption from income tax payments) to investment tax allowances, abatements for exports, and deductions for R&D training. The only major requirement for taking advantage of these incentives is that the participating company's operations must include a significant degree of participation of the Bumiputra (Malaysian indigenous peoples) in order for the company to be listed on the stock exchange.

In addition to the wide variety of tax and investment measures Malaysia is using to attract investment and generate local capital formation, it offers monetary regulations that provide a relatively free flow of money into and out of the country, particularly for designated projects. This allows MNCs and other outside investors, such as Japanese or U.S. companies, to invest their capital and to transfer funds as needed with a minimum amount of government involvement. These measures also benefit local companies seeking to invest in the region.

In terms of its regional investments, Malaysia, like Singapore, increasingly emphasizes interregional technology flow, the goals being to develop the simpler economies of the region as markets and as sources of lower-end products. Malaysia's additional goals, as in Singapore, are to develop joint operations, such as R&D centers and high-technology parks, and to pursue resolution of such regional problems as those of maintaining stable water and power supplies. Malaysia provides a significant percentage of Singapore's support and lower-end work force.

All of the various tax incentives, joint projects, and the like that Malaysia offers combine to make the country a very attractive investment site. Japan, the United States, and other countries (primarily Taiwan, Singapore, Hong Kong, and Germany) invest heavily in the country — RM$15 billion in 1995. Such incentives as Malaysia offers of course help develop the country's access to new technology, help support its own R&D efforts, help train a better skilled work force, and in general help raise the living standard of the entire population. These measures are an integral part of Malaysia's previous, current, and future development plans, especially those encapsulated in *Vision 2020*, a developmental model very similar to that of Singapore and the other East Asian mini-dragons. A clear case in point is the emphasis on the part of the Malaysian government (particularly through the MTDC) and the Malaysian educational institutions on an increasingly close relationship between foreign and local industries.

One major infrastructural advantage that Malaysia has but Singapore and the other mini-dragons do not have is its preferential status under the Generalized System of Preferences (GSP) in first the GATT system and now the WTO. GSP status grants special trading privileges and rights in multilateral trade negotiations that provide special access to the First World countries' markets and industry protectionist measures. These privileges give Malaysia a critical advantage in the world marketplace and give its local companies a significant competitive edge over their international competitors. An important caveat here is that in the WTO accord of 1992, there are provisions and tentative agreements to deal with cases like Malaysia where a country's GSP preferences stretch the intention of the GSP system as it was first agreed upon by GATT members in the early 1970s. The basic intent of the new WTO agreements is to decide at what point a developing country's economy is no longer so fragile as to warrant its being allowed continued preferences under the GSP system. Malaysia's GSP status may undergo some review under the 1992 WTO provisions.

7.2.2 Role of Industry - Specifically Electronics

Central to Malaysia's development plan is the major role to be played by its electronics industry. Malaysia is already one of the world's leading exporters of electronic components, and it is trying to eventually attain a capability of item-complete production, because this significantly increases the value added. As a current major producer of ICs, semiconductors, capacitors, and wafers, Malaysia's electronics group grew nearly 29% in 1995, following growth of 21.6% in 1994. The construction of its FTZs underlines this growth pattern focused on electronics: seven of eleven of them emphasize electronics. Another example is a major project in Kulim Hi-tech Park that will emphasize the production of software, ICs, and wafers, including multilayer, coextruded, electron-beam cross-linked polyolefins, shrink films, and HDD (to be 100%-financed by U.S. investment). Further, Malaysia has more than 80 companies producing PC-based components and/or subassemblies and peripherals. It has also designated HDD as a strategic industry with a 10-year tax holiday. To complement this industry, the Malaysian government has undertaken to develop both upstream and downstream industries to produce components such as motherboards, SCs, wafers, and software.

Malaysia is also the world's number three producer of and leading exporter of finished semiconductor chip devices. Its 1995 semiconductor output totaled approximately $8 billion and accounted for over 40% of Malaysia's electronics industry. Discrete semiconductors and integrated circuit packaging hold the number one position for Malaysian electronic production, accounting for over 25% of total electronics production in 1995. InterConnect Technology, for example, founded by the Malaysian government, is the country's first state-of-the-art semiconductor manufacturing facility. Its mission is to produce semiconductor products for customers worldwide. It currently produces 25,000 8-in wafers per month, is equipped with 0.35 μm equipment, and has a plan to achieve 0.25 μm.

The Malaysian Institute for Microelectronics Systems is another example of the already excellent and rapidly improving status of the country's electronics manufacturing system. It is a microchip plant being built by the government for $44 million that will be fully functional in 1996. The government will also set-up a microelectronics center in Kulim Hi-Tech Industrial Park by 1997.

There are three major points to stress about Malaysia's future course of growth and development, with special emphasis on development of its electronics industry:

1. Like Singapore's plan E 2000, Malaysia's Vision 2020 plan emphasizes

the movement of Malaysian electronics manufacturing to more advanced production capacities and diversity.
2. The plan places increased emphasis on greater and more advanced technological development
3. Likewise, the plan calls for increased development of all of its infrastructural components.

7.2.2 Future Plans

Over the last eight years, driven largely by manufacturing, Malaysia's GDP has averaged an annual growth of 8.5% with a real GDP growth of 9.6%. It is shooting for a continued GDP growth of 8.5% (global trade is expected continue to expand at 8%). Malaysia's growth is based on clear plans and coordinated activities to maximize its capabilities and reduce or eliminate its weaknesses to achieve its development goals. Its broadest vision is enunciated in *Vision 2020*, and its implementation plans are detailed in its 7th Development Plan. This plan includes such provisions as capital investment of $800 million in Intel's Penang facility. Infrastructural actions include those in education (including technical institutes); trade-sale and/or marketing; telecommunications (fiber, wireless, etc.,); mass transit; importation of technical experts; the development of technical associations and technical consortia (e.g., ball grid array), and importation of laborers from such countries as Bangladesh, Indonesia, Philippines. This last provision is due to its tight labor market — its unemployment rate is less than 2.8%. Increasingly, Malaysia is exporting lower-level production to other countries and importing workers, although in the longer term, it plans is to reduce dependency on these workers. It is following the same basic strategy as Singapore to increase the educational level of its people as a means to developing a capability for higher value-added production.

A key portion of the 7th Malaysia Plan is the OPP2 (Second Outline Perspective Plan) for 1991-2000. In this plan, the Malaysian government targets the manufacturing sector as the major engine of growth to achieve the Vision 2020 goal of attaining developed nation status by 2020. The interim goals are for the manufacturing sector to grow by 10.5% per year and contribute 37.2% to GDP and 82% to total export earnings by the year 2000. Other more general targets are to create and sustain an environment that will attract and hold continuing FDI; to increase domestic investments; to continue to diversify its economic base; to continue to develop and advance technological innovations; and to continue to develop R&D focus in education, training, and technological centers. To pursue these goals, the country will attempt to perform several actions:

- reduce the corporate tax from 32% to 30%
- allow a full tax exemption on all income earned abroad and reinvested in Malaysia

- allow a reduction in the withholding tax on interest payments made to nonresidents from 20% to 15%
- reduce the tax on technical fees and royalties from 15% to 10%
- reduce import duties on a wide variety of more than 2,600 items
- abolish sales tax on machinery parts and components and imported heavy machinery

Under the Seventh Plan, Malaysia will develop itself as an R&D center. Under the Action Plan for Industrial Technology Development, as part of its overall Vision 2020 plan, Malaysia is taking the following steps to enhance the country's R&D capabilities:

- set up the "Intensification of Research in Priority Areas" scheme to link the public R&D activities in all strategic areas to the country's needs
- set up the Malaysian Technology Development Corporation (MTDC) to provide seed capital to companies interested in commercializing the research results of research organizations and universities
- establish the Technology Park of Malaysia to accelerate development of local R&D information technology, biotechnology, electronics, microelectronics, resources-based technology, telecommunication technology, aerospace, defense, and material technology; also, set up the Kulim Hi-tech Park and the Johore Technology Park
- develop R&D incentives to increase the pioneer status allowances to 5 years and investment tax allowance to 100% on qualifying capital expenditures; also, provide single deduction eligibility for revenue expenditures incurred in R&D
- provide matching grants and industrial building allowances of 10% initially and 2% annually
- provide capital allowances and tax exemptions on machinery and equipment for R&D, and provide a relaxation of tax conditions for expatriate and foreign workers

Combined, these measures underline Malaysia's commitment to future growth and development. The four major goals of this development are as follows:

1. develop greater cooperation between industry, government, and the educational system to optimize the country's GDP growth
2. increase emphasis on the development of technology and innovation
3. persuade other ASEAN countries to continue reinvesting in core competencies at all levels of the production chain
4. develop the country as a major force in the world electronics market.

7.3 SUMMARY

Singapore has one of East Asia's widest ranges of government programs to support the development of domestic industry, the electronics industry in particular. The country has increasingly sophisticated infrastructural, organizational, technical, and manpower capabilities. In order to foster the rapid growth of Singaporean industry, the government is aggressively attracting foreign industry and direct foreign investment to the country. It seeks to attract MNCs to establish Singapore as their East Asian center of operations; invest in not only Singapore but its surrounding region by establishing an equity stake in their development; and invest in and transfer leading-edge and higher value-added technology. Fundamental to all developmental efforts is the close-knit and very effective interrelationship between the government, the universities, and industry.

Malaysia has the basic infrastructure, growing technological expertise, and expanding manufacturing diversity to meet the demands of its customers in terms of the quality and quantity of products it can produce, but it also has the very important trade and development advantages granted it under the General System of Preferences (GSP). Malaysia's infrastructural development is not as sophisticated or developed as Singapore's; however, it is striving in that direction, and it currently does have trading advantages that Singapore does not enjoy.

In summation, Singapore and Malaysia already command a major share of the world's electronics markets. With their respective detailed and visionary development plans (E 2000 and Vision 2020) they are seeking to command an even larger portion, particularly in the end-product sectors. Their capability to continue to expand economically hinges to a great extent on this. They are accelerating regional development with increasing technology transfer, cooperation agreements and other activities, and this action benefits their further individual development. Each of these two Southeast Asian nations are developing a cadre of workers within a focused, driven society, who promise to provide them the human resources to achieve these aims. They are absorbing and further developing the results of international R&D efforts and exploiting these efforts to further gain their developmental goals. Their immediate and long-term investments in facilities and capital also supports this pursuit. If past success is any indicator, it is likely that each of these two countries will seize increasingly large shares of the world's electronics markets, and hence command increasing influence in the world economy as a whole — influence disproportionate with their size.

REFERENCES

Abdul Hamid, Ahmad Sarji. *Vision 2020*. Selangor Darul Ehsan, Malaysia: Pelanduk Publications (1995).

Beh, Dr. Swan-Jin, Director, Singapore EDB Office, Washington D.C. Conversations with authors, as well as review of draft text (September 1996).

De Scuza, Robert, "Singapore: Strategic thrusts for competitive success in the electronics industry," Journal of Electronics Manufacturing (1994).

Dunn, P. "Malaysia is poised for leap into semiconductor production." *Solid State Technology* 52ff (December 1995).

Economic Development Board of Singapore (EDB). *Electronics 2000: Industry Report* (May 1995a).

EDB. "Singapore's electronics R&D in full steam." *Singapore Investment News* (July 1995b).

Economic Development Board (EDB), Manufacturing 2000: The cluster development approach (1996).

Elsevier Advanced Technology. *Semiconductors: Profile of the World Semiconductor Industry - Market Prospects to 1999.* . Oxford, United Kingdom, Elsevier Science Ltd. (1995).

Elsevier Advanced Technology. *Yearbook of World Electronics Data 1996, America, Japan, & Asia Pacific*. Vol. 2. . Oxford, United Kingdom, Elsevier Science Ltd. (1996)

Integrated Circuits International (ICI). "NS invests in Singapore plant." (March 1995a).

ICI. "Singapore invests in wafer production." (April 1995b)

ICI. "Memory joint ventures in Malaysia and Singapore." (March 1996a)

ICI. "Semiconductor equipment orders." (March 1996b)

National Survey of R&D in Singapore, 1994, National Science and Technology Board Singapore, Science Park, Singapore, 1995

Shariffadeen, T.M.A., "Malaysian Electronics Industry — Transition and Transformation," IMS 93 keynote address, August 1993. Proceedings of the International Microelectronics and Systems Conference 93.

INDEX

Association of Southeast Asian Nations (ASEAN), 4, 40
Centers of Excellence, 58
companies, 45
Current State of Development, 93
Economic Development Board, 47
EDB, 47
Electronics in Malaysia, 39
Electronics in Singapore, 16
ELECTRONICS PRODUCTION IN MALAYSIA, 63
ELECTRONICS PRODUCTION IN SINGAPORE, 45
Financial Incentives, 77
Future Plans, 91
GATT, 29
General Agreement on Tariffs and Trades, 29
GINTIC Institute of Manufacturing, 21, 59, 73, 76
Government Support, 72
IMP, 35

Industrial Master Plan, 35
Institute of Microelectronics (IME), 58, 73, 74, 75
investments in semiconductor manufacturing, 63
Malaysian Institute of Microelectronic Systems, 68
Manufacturing in Singapore, 13
MIMOS, 68
multinational corporation, 45
National Science and Technology Board, 53
NSTB, 53
People's Action Party (PAP), 2
trade, 1, 4, 29, 40, 41, 42, 66, 73, 79
United Malay Nationalist Organization (UMNO), 2
University Support, 74
Vision 2020, 72
World Trade Organization (WTO), 4